T0229879

ENGINEERING DOCUMENTATION FOR CAD/CAM APPLICATIONS

MECHANICAL ENGINEERING

A Series of Textbooks and Reference Books

EDITORS

L. L. FAULKNER

Department of Mechanical Engineering
The Ohio State University
Columbus, Ohio

S. B. MENKES

Department of Mechanical Engineering
The City College of the
City University of New York
New York, New York

1. Spring Designer's Handbook, *by Harold Carlson*
2. Computer-Aided Graphics and Design, *by Daniel L. Ryan*
3. Lubrication Fundamentals, *by J. George Wills*
4. Solar Engineering for Domestic Buildings, *by William A. Himmelman*
5. Applied Engineering Mechanics: Statics and Dynamics, *by G. Boothroyd and C. Poli*
6. Centrifugal Pump Clinic, *by Igor J. Karassik*
7. Computer-Aided Kinetics for Machine Design, *by Daniel L. Ryan*
8. Plastics Products Design Handbook, Part A: Materials and Components; Part B: Processes and Design for Processes, *edited by Edward Miller*
9. Turbomachinery: Basic Theory and Applications, *by Earl Logan, Jr.*
10. Vibrations of Shells and Plates, *by Werner Soedel*
11. Flat and Corrugated Diaphragm Design Handbook, *by Mario Di Giovanni*
12. Practical Stress Analysis in Engineering Design, *by Alexander Blake*
13. An Introduction to the Design and Behavior of Bolted Joints, *by John H. Bickford*
14. Optimal Engineering Design: Principles and Applications, *by James N. Siddall*
15. Spring Manufacturing Handbook, *by Harold Carlson*

OTHER VOLUMES IN PREPARATION

ENGINEERING DOCUMENTATION FOR CAD/CAM APPLICATIONS

CHARLES S. KNOX

Knox Consulting Services
St. Paul, Minnesota

MARCEL DEKKER, INC.   New York • Basel

Library of Congess Cataloging in Publication Data

Knox, Charles S., [date]
    Engineering documentation for CAD/CAM Applications.

    (Mechanical engineering : 30)
    Includes index.
    1. CAD/CAM applications--Documentation. I. Title.
II. Series.
TS155.6.K67   1984   620'.0043'02854   84-1773
ISBN 0-8247-7089-7

MARCEL DEKKER, INC.
270 Madison Avenue, New York, New York 10016

Current printing (last digit):
10 9 8 7 6 5 4 3 2 1

PRINTED IN THE UNITED STATES OF AMERICA

To my two sons, David Stuart and Gary Sinclair, with love and respect

PREFACE

The purpose of this book is to emphasize to engineering, manufacturing, and accounting the importance of consistent, well-planned, and computer-oriented engineering documentation systems. That such systems are needed to optimize flow of information and to increase the efficiency of modern CAD/CAM systems will become apparent as their interrelationships are discussed. Too many times engineering departments give little thought to the documentation and communication requirements of other departments. Thus, this book emphasizes and explains a set of principles and procedures needed to improve these requirements.

The word "flow" conveys an important concept when used with engineering documentation. Engineers consider their main task to be designing "things." Too little time and effort has been spent by them to streamline methods for conveying the data they generate which "flows" from engineering through manufacturing to accounting. By "flow" we mean the means by which engineering drawings, their associated data, bills of material, engineering changes, part number or revision level changes, parts lists, and geometry are documented and disseminated to all other departments.

With the advent of CAD/CAM and the emphasis on integrated computer systems, which now also include new data such as design geometry, the need for a more consistent flow path has become mandatory. After all the segments of engineering documentation have been discussed, the actual flow of data will be described so the reader can sense the reasons for each of these segments and their importance.

Some of the subjects we will cover include a simple part number system which conforms to rules of interchangeability and configuration management, piece part drawings for flexibility of design retrieval and manufacturing planning, design retrieval to minimize the proliferation of new designs, a mechanized parts list and bill of material for communication and integration of planning data with manufacturing systems, product structures for conveying manufacturing assembly requirements, and modular design to minimize customizing bills of material for each new order.

Of course, many companies have well-documented systems which operate efficiently today. For those companies, this book will serve to confirm their philosophies. Many more, however, do not have efficient engineering documentation systems. Long part numbers with obsolete or unimportant significance, multiple parts on one drawing, customer job cost control, and many more inefficient and wasteful systems still exist. This book will help those companies who have such problems to prepare plans for change, and it sets forth some of the rules and principles for making changes accurately and efficiently. It should be considered as a guide for any company whose management has decided to increase the efficiency and effectiveness of their data processing systems, to implement CAD/CAM, or to increase the productivity of their engineering department.

Consequently, this book has been written for engineering managers such as vice presidents, chief engineers, engineering administrators, managers of engineering standards, and their staffs. Since the change in engineering documentation will influence manufacturing systems, other readers who will be assisted include the vice president of manufacturing, manager of manufacturing engineering, production control manager, inventory control manager, purchasing manager, quality control manager, and their staffs.

Since accounting personnel, too, will be affected by changed engineering documentation, they will profit from recognizing the benefits derived by the methods described here, particularly in obtaining more accurate costs and estimates in using a bill of material system.

Data processing personnel and the corresponding data base manager can utilize the information found in this book in planning and implementing business systems. Last, but not least, this book will provide an important dimension to an engineering student who wants to contribute his skills to a new employer more rapidly.

The content of the book has been planned to take the reader from the simple to the complex, from individual items to interrelationships. In this manner, we hope the reader will find it comprehensive but clear, all encompassing but understandable. The productivity of manufacturing companies relies so much on improvement of the systems you will find in this book. We hope they are beneficial to you.

Our appreciation and thanks to American Hoist and Derrick, Control Data Corporation, Anaren Microwave, Inc., and Ingersoll-Rand, Turboproducts Division, and Conserv Corporation, Mendota Heights, Minnesota.

Charles S. Knox

CONTENTS

INTRODUCTION

1

A friendly revolution has begun in engineering departments of almost all manufacturing companies. Over the years, design methods have gradually been changing. More sophisticated methods of determining optimum designs have been invented, and competition, both foreign and domestic, has made it inevitable that the engineer accept and utilize them. The computer has become a useful and well-understood tool in this endeavor.

Today, for example, it is quite usual to design gear sets by a gear analysis program. No longer does an engineer guess at a safety factor in some structural design, but, rather, uses a program such as NASTRAN or Stardyne to examine the displacements and forces on the product using finite element analysis and plotters to draw the stress analysis results. Guesswork and "cut-and-try" attitudes are eliminated.

In the 1960s, computers such as the IBM 704 as well as other smaller computers were fully appreciated and utilized by engineers, particularly in the aerospace industry. Many companies, however, felt there was no significant economic advantage to using computers to perform design calculations. Eventually, however, remote terminals, used for very limited applications, were installed even in these "holdouts."

The result of this slow but steady inroad of computers for design applications did not, unfortunately, cause the engineer to see and understand the significance of such calculations as a means to transmit the resulting data "electromagically" to other departments of a company. Consequently, the results of most design calculations were arduously recopied onto appropriate drawings or bills of material.

The "paper work" of the engineering department languished. Often a designer or drafter, too slow or less competent, was transferred to the engineering standards department. The standards thus generated did not portray the future requirements, but merely documented the past. Seldom was an attempt made to determine whether practices, present or past, were beneficial and economical to other departments. Not infrequently a contest emerged between engineering and manufacturing. Manufacturing personnel stated, "I can make anything they can design." Engineers took up the challenge raised by manufacturing, creating ever more exotic designs which were not always necessary. Thus the battle was joined.

Another phenomena also had its effect. Many companies used to place their en-

1

gineers into training programs which allowed them to work in, and become acquainted with, various manufacturing disciplines. It was not unusual to see graduate engineers working on an assembly line as a part of this training. As engineers became scarce, and as union rules or manufacturing practices discouraged this direct participation, the engineers slowly lost touch and became unaware of the manufacturing techniques, their availability, and their limitations. To this day, many engineers design products, parts, and assemblies, without the slightest idea of how, when, or where they will be manufactured. In addition, the long lead time between the finished drawing release and the first part made is not conducive to feedback in assisting the engineer to understand the errors may have been made. Lack of this feedback loop has helped to perpetuate errors, more costly designs, and a lack of awareness by engineers of the problems they may have contributed to the manufacture, product cost, and increased lead-time requirements.

Engineers are rarely given cost data. Some chief engineers feel such knowledge might destroy the designers creativity. Rarely does any design analysis program utilize cost as a variable. Consequently another type of feedback to the engineer is missing. Value engineering has been useful, but is sporadic.

In much the same manner, the influence of data processing and data base requirements of a manufacturing firm have not been emphasized to engineers and designers. If hundreds of people in manufacturing must write 43-digit part numbers every day for a host of parts, the engineer may be blithely unaware of the consequences in time, accuracy, and confusion. If the engineer prepares a handwritten bill of material on a drawing, and someone else retypes it into a form for a computer system, the potential errors do not seem to affect him or her.

Aerospace and computer manufacturers learned early the penalties of poor data generation. The complexity of the product simply did not allow for many errors to occur. Giant steps were taken by such companies to clean up their data so it could be processed in a streamlined manner. Unfortunately any progress on their part could be easily sneered at by other chief engineers. After all, "the government" was paying for this fancy kind of documentation. The realization that economics of scale and faster flow of data were achieved seemed to go over their head.

In addition to aerospace and computer manufacturers, manufacturers in the automotive and tractor areas adopted sound engineering practices and good documentation control, short part numbers, and some retrieval systems. While these documentation techniques were well planned, they were essentially manual. Conversion of these methods to a more mechanized process came only after a long and agonizing period of adjustment.

Today, two basic types of documentation systems appear to predominate in the engineering field. The first could be described as a job order system. In this system the engineering documentation methods are almost completely ignored by manufacturing and accounting. The part number becomes secondary to an accounting job number which is tied to a specific customer order. Engineering changes are meticulously documented according to the customer affected. Interchangeability is carelessly applied. The motto "we never make the same thing twice" is gleefully and proudly proclaimed.

Personnel of a company manufacturing rock crushers stated, "no two rock crushers are identical." The discovery that 90% of all the parts in every rock crusher were identical did not cause the slightest change in the job number system, which required that each part of each crusher be manufactured under a separate job number, completely ignoring any advantages of mass production. The lack of awareness on the part of both engineers and accountants in this type of system has contributed to unnecessarily high cost products, and, in several cases, eventual bankruptcy.

The second type of system could be described as a part number system, and presumes a part or a product can and will be used again. Thus, the part number documentation is used not only in engineering, but also in manufacturing and accounting. Many companies are moving to develop systems such as this; few, if any, are converting to a job number system.

No matter which of these systems has been in existence within a company, both types of companies have resisted mechanization. Let's examine some possible reasons why this resistance persists.

We have already described how the computer has been used in design but the results have been manually transferred to other media such as an engineering drawing. Perhaps the simplest answer to the resistance to mechanization can be found in the attitudes, the education, and the objectives of the engineer and designer. Engineers design "things." These "things" do not portray "flow." The design is an entity, not a portion of a stream. The result of a design is a picture, not a piece of paper. The change in a design is a new picture, not an engineering change notice. The parts composing the design are additional drawings, not physical parts made with corresponding tools, work centers, and inventories.

Engineers do not receive data, they create it. Therefore, they have never felt the urgency of receiving late, incomplete, or inaccurate data. They have been shielded from the consequences of their actions. One example might illustrate this condition.

A manager of production in a company which manufactures tractors noted that a hole cover was designed to be chained to the body of the product; the chain was fastened to the cover by means of a hole through a block, which was, in turn, welded to the cover. The block had been welded off-center to the cover. If the block could be moved to the center of the cover the cover could be machined more rapidly and economically using the block in the chuck of the machine. The present design required holding the cover in "soft jaws" to machine it. A call to engineering was met with the comment, "You don't understand what we are doing, just build the parts as we have designed them." For 30 years the production manager watched while six covers per product were manufactured expensively.

The engineer must become a part of the engineering–manufacturing process. Competition, economics, the very existence of a company mandates this. Isolation must cease. It is time to examine the means by which engineers and designers communicate to the rest of the world.

At times data processing personnel have made some tentative attempts to try to rectify engineering systems. They were outnumbered. Unfortunately, too, data processing personnel are usually more accounting than engineering oriented. Thus it was difficult for them to have empathy for the problems in engineering and to find the means to solve them. Only engineers can completely solve engineering problems. We must elicit their interest and cooperation.

If nothing else has been able to induce engineering systems to change, computer-assisted drawing/computer-assisted manufacturing (CAD/CAM) will. Perhaps at first CAD/CAM systems will be installed to perform drafting. If this occurs it will be unfortunate, since the full power of an integrated CAD/CAM system will not then be utilized. Soon, however, some uneasiness is bound to appear when either drafting productivity does not meet expectations, or the speed with which drawings are prepared is lost when other data is transmitted in the old and slow manner. CAD/CAM may well be the major reason why engineering systems must be changed and the reason they will be changed. Let us then look at the outline of a flow of data as it may happen when an integrated CAD/CAM system is utilized. We will then point out those sections in this book which we will discuss that will assist in streamlining the documentation flow.

Figure 1 illustrates data flow through engineering as marketing and engineering cooperate to prepare a new product or utilize an old one. The chart implies a cathode ray tube (CRT) on-line to a data base which contains all current product specifications and the corresponding bills of material. Entering this data base with customer specifications allows the engineer to obtain a family of products which have been previously designed to meet the general requirements of the customer. Examining the available products and comparing them to the customer specifications may determine a match on some or all of the modules of an existing product. If such is the case, the proper bills of material can be immediately identified and issued. We will discuss how this type of data can be planned in the section on "Modular Design (Chap. 9).

Should no existing product meet the requirements, a "family" bill of material and set of specifications can be displayed. A new design can be initiated utilizing preplanned design specifications, design analysis programs, and a family bill of material. We will discuss how this data might be prepared in Chapter 5 on design retrieval and family drawings for CAD/CAM.

As the general bill of material is further detailed, use of the family of parts implied by the bill of material can be used to prepare a cost estimate. In addition, a skeleton bill of material can be prepared and, with the family lead times, used to develop a "Christmas tree," or exploded bill of material which can be used to graphically portray when each new design must be completed. Any parts which can be utilized from old designs can be "released" into the normal stream so that planning for their acquisition may be done immediately through the material requirements planning system (MRP). We will discuss the procedures necessary to achieve this in Chapter 7. Finally a customer proposal can be prepared with system drawings, delivery dates, and costs.

To make this possible isn't it important to make sure the engineering systems become part of the company flow and that they are streamlined?

Figure 2 shows a similar flow chart through a CAD/CAM system at the piece part and assembly level.

When a design requirement is known, the part or assembly is searched for through a design retrieval system. A family drawing and a corresponding catalog with critical dimensions are displayed. Should the dimensions for a previously designed part appear to be correct, the actual piece part drawing is displayed for confirmation. If it is adequate, it may be released immediately. If not, the family drawing is then recalled and the alphabetical dimensions replaced with the actual dimensions. If necessary, design analysis techniques can be used to optimize the design, and producibility tips from industrial engineering applied to minimize the cost of the new parts. The family drawing and the finished drawing are prepared using customized design and drafting standards. We will discuss the economic advantages of piece part drawings later.

Once the design is complete, the "text" data can be "stripped" from the design and transmitted to the data processing business systems requiring such data.

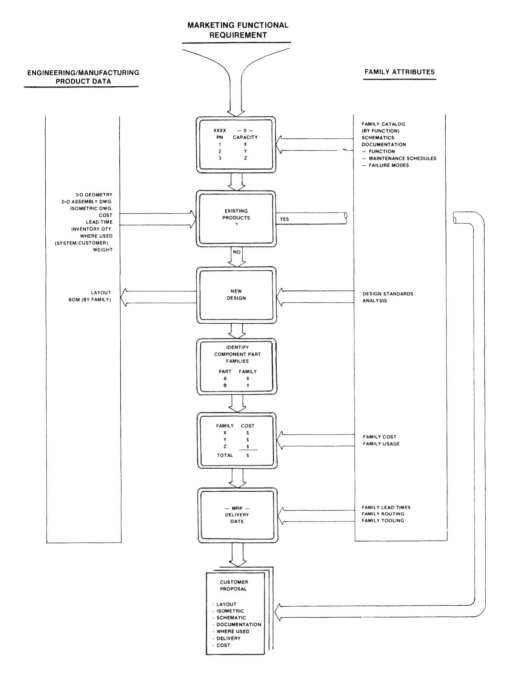

Figure 1 Integration of CAD/CAM systems in marketing and engineering require that engineering documentation have integrity and an orderly flow.

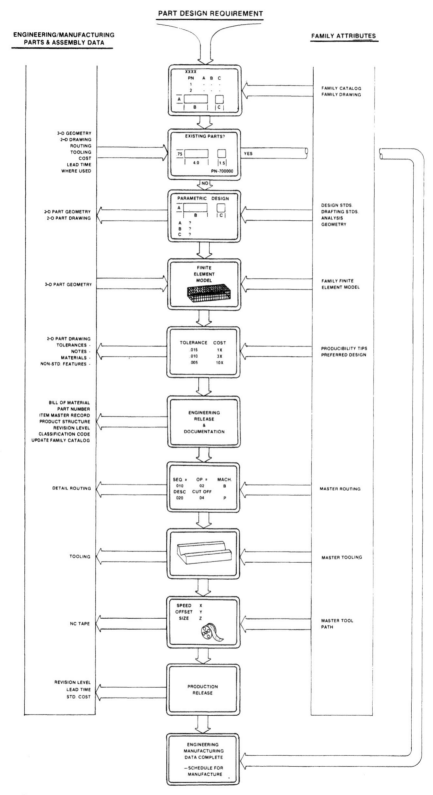

Figure 2 Detailed design requires an accurate data base, a retrieval system, and accurate engineering data.

This data is also discussed in Chapter 7. The need for proper structuring of the data is discussed in Chapter 8. Finally, engineering may use the geometry of the new part to prepare a specific routing based on a family master routing. In addition, tooling and numerically controlled (NC) machine tool cutter center line data can be prepared from the original geometry and surfaces prepared by the designer.

If we are to take advantage of this powerful tool, the ingredients must work together not in opposition. We hope the following chapters will go far to clarify the reasons for change and introduce the means for so doing.

A WORD ABOUT
DATA PROCESSING
2

For many reasons, engineers and designers have looked on data processing as an unnecessary evil. "Let someone else do it" has been the motto. Consequently, clerks have been enlisted to take basic design data retranslate it into the information required for business systems. Without participating in the design of such systems, the engineer does not recognize the faults or inconveniences of the old engineering systems which are being used. Perhaps some brief analogies and a comparison of past communication devices will serve to illustrate and emphasize the role data processing plays in engineering specifically, and with a manufacturing company generally.

An excellent analogy which can be used to define data processing is the action of the human body. Every day we obtain, as "input," a large variety of dissimilar food and drink. All of this input is sorted, classified, and digested, and then disseminated through the body where blood, tissue, bones, and waste are manufactured. From other sources we obtain sounds, pictures, and odors. The integration of these various sensations cause our body to react physically, or our mind to respond internally, or in combination. This can be termed "output." If we compare this process to business, we can see that sales orders, engineering drawings, sales forecasts, and parts requirements are input to a business. These factors which make a business run must be digested, integrated, and disseminated until the proper output, a physical product of manufacturing is forthcoming.

The skill and speed with which information flows determines the efficiency which results. A sluggish human being cannot operate efficiently. Neither can a sluggish business.

Two of the greatest drawbacks to the use of present day data processing is the inability of the management of a company to see:

Dissimilar Information Can Be Made To Flow Smoothly Through A Company
Dissimilar Information Must Be Made To Flow Through A Company Smoothly

We are in the midst of an industrial revolution second to none in the history of the world. We must either prepare for it or not survive the competition of those who do. The keywords are *flexibility* and *speed*. Very often those who discredit data processing do so under the guise of "conformity." They honestly believe mechanized processing of records relinquishes the right of the worker to think. It is possible to prove this concept is wrong. Let us use an example to do so.

Through the years an inventory control department has been figuring gross and net requirements of parts and assemblies manually. As the business grew it was necessary to add more and more people to perform this job. Standards were determined to define how many parts each person could handle. Today these people are still working under the program designed several years ago; it has merely grown larger.

When the system was first devised, no analysis of economic lot sizes, no combining of similar items, no mass engineering change dates, no controlled lead times, no inventory ratios, and no inventory classifications by value were required. Of course, if they had been required, the size of the operation was such that a matter of two or three hours would have accomplished the study. Many decisions affecting these problems were done "instinctively" by people who knew the product, had memorized the part numbers, and "guessed" very effectively. It was not usually the clerk at the inventory control desk who made these decisions. It was the boss who took the computations made by the clerk and altered them to fit his or her instincts. The clerk was always a clerk. A clever clerk learned to read the implications behind the boss's decisions. An alert clerk who wanted to "get ahead," developed an "instinct" to make decisions. Since sound reasons were not always given for the decisions made, changing conditions did not always reflect changing rules. Nevertheless, these instincts very often became "rules" and "procedures" which, while sometimes effective, could also be disastrous.

As various sophisticated considerations had to be dealt with, such as economic lot sizes, and combining orders, the procedures were often compartmentalized into segments which one person could handle easily. Not realizing the full implication of such a division, the procedures thus installed could handle only one variable at a time.

Very often two people, working in the group determining the economic lot size to purchase for a particular piece part, would have found they were both wrong had the variables being considered been combined. Furthermore, and most important, it very often took far longer to get the data needed to make such a study than it was worth, or that time allowed. As a result, decisions, although better than had these factors not been considered, left much to be desired.

Another problem was created as more considerations became necessary in such an operation as inventory control. Each time another factor was considered, a new piece of paper, a new procedure, a new flow was tied onto the old ones creating a bulky, clumsy, complicated flow of information. Each department assimilated the new information in its own way, and very rarely were departments compatible.

Where then does data processing fit in this picture? First we must make one fact perfectly clear. Even without data processing, tremendous advances in speed and flexibility could be obtained by standardizing information flowing through the company. This, in essence, says most companies today are really lagging behind normal modernization. Departmental barriers, lack of foresight, and failure to create projects or task forces to study the entire flow have resulted in this lack of progress.

We could never understand fully the movement and flow and strategy of a football team if we were always on ground level. Only when we are able to stand off and look down at the whole game does it become understandable. We must apply the same technique in studying the flow of information in a company.

There is another concept in football that merits consideration. The quarterback calls the signals based on his analysis of the best way to move the ball. The rest of the team may not see the logic for the plays called. Furthermore, if all plays continually force the left linebacker to make a supreme effort, he is liable to find reasons why the plays called are wrong. Admittedly, too, each of the 11 members of the team could call signals as easily as the quarterback. However, a "committee" approach in the huddle would certainly cause "delay of the game" penalties and, no doubt, move the ball backward rather than forward. To make a touchdown requires strategy, a planned program, and a cooperative, educated team, as well as strong leadership.

Let us assume one department of a company is modern in its concepts and has a truly integrated flow of information. Ideally, every other department should follow this pattern. Obviously this would require many changes for many people. Education alone is not the answer. The people who must change are "on the football field," and only confidence in their superiors can keep them from fearing the results. They, in turn, cannot judge the merits of the new procedures. Consequently, we have determined the two prerequisites for developing new and modern systems:

Administrative Approval and Assurance is Mandatory

Changes Will Affect Every Department and Must be Explained

Our example of only one department being modern is rare. Even if it were true, the integration of all departments will cause some changes. They, too, must be prepared.

Inevitably standardizing procedures results in one department obtaining an increased work load. This is usually because the department which has been performing some particular function has done so to obtain data they need, whereas another department is actually the source of the information. This will be particularly true as engineering systems become modernized, as many tasks which engineering departments should initiate have become, through abdication of responsibility, the work effort of others. Such changes, if not backed by administrative approval of the entire program will cause eventual breakdown of the project.

If we were given the task of redesigning a particular product we sell, there can be no restriction which says, "don't change the design of component A." The component we cannot change may make the redesign impossible or severely curtailed. Freedom of design based on a sound study is a "must" also for data processing.

Let us return to our inventory control clerk's fear of "conformity." For years the clerk has "conformed" to procedures. Otherwise, chaos would have been the result. They are, however, familiar conformities. Once our inventory control clerk has been given new and standard tools to work with, it will be clear that dull, repetitious, work has been delegated to a machine. The clerk, too, can now "stand back" and analyze on a sound basis the results of his or her efforts. Familiarity will, once again, help the clerk to "conform." However, the results obtained will be documented on a sound, accurate basis, very often presented as a series of choices calling on the clerk to achieve ever higher technical ability. This employee will be released from the status of "clerk" or "recorder" and allowed to play a more active and useful part in the function of the company.

Another word about "standard tools" is important. If a person is given a hammer and told that he or she will eventually be able to make a fine piece of furniture, they may become very skeptical. They may not only be unable to use the hammer, they are also aware that other tools are needed. Describing tools that will be available in the future does not always make them feel the need for learning how to use the hammer skillfully. Those advocates of data processing should learn not only how to show the need for using the hammer, but also how to describe future tools so that their potential is understandable.

On the other hand, the recipients of these tools should learn to accept the credibility of the information given them and strive to achieve the skills they will need. Only this two-way cooperation can be effective.

A mechanized system cannot perform miracles. It can only help. The people in an organization must be integrated first before the full usefulness of a computer is felt. That is why computers today are secondary in importance when considering data processing.

Finally, let us discuss one more important aspect of data processing which has created fear in the minds of many people including engineers, designers, and drafters: The need for less people when a system is mechanized. Unfortunately, computer manufacturers and companies who use computers have many times justified their use by attempting to show how fewer people are needed as data flow is mechanized. While such justification is true in most cases, these advocates have ignored two important end results:

1. There is a need for more people as a result of mechanization.
2. The inability to obtain sufficient people has required mechanization.

Today most insurance companies, for example, are mechanized. The number of people employed in this field, however, is approximately the same or higher than before. The economics of mechanization have been passed on to those requiring insurance, making it available and economical to more people. Just as mass production lowered the cost of automobiles so more people could purchase them, so has mechanization enabled more people to buy more insurance. Mechanization, just as mass production, increased productivity. The increases in productivity lower costs and increase demand, which in turn creates more not fewer jobs.

Conversely, mechanization has augmented our labor market. If, suddenly, no mechanization was available, there would not be enough people in the United States to fill the vacancies. We would be floundering in paper work and tedious manual computations which would be almost impossible to achieve. The necessity for doing such computations manually would slow to a crawl our scientific and cultural progress. Progress in some areas would cease altogether.

As we approach the discussion of modernizing engineering systems and integrating them into the flow of the entire company, we must be aware of the philosophies discussed in this chapter. Some recommended changes may be painful. For example, a long established part number system in a company is not altered lightly. Changes to systems create a lot of emotional reaction if the changes are not properly planned, documented, and explained. We hope the sections on some of the major problems in engineering systems will assist in studying potential changes objectively.

In the last few years, cathode ray tubes for entry of data into a business system have been introduced into engineering departments. All sorts of reasons and excuses for people other than engineers and designers to use them have been presented. It is time to stop this nonsense and show that the most timely, accurate, and logical procedure for generating data for business systems is when the engineer and designer enter their own. There is a suspicion the engineer feels demoted when using such a device.

The purpose, then, of mechanization is threefold:

1. Mechanizing the operations of a company is designed to take little information used by all people or all departments and convey it without the need for initiating repetitious documents.

2. Mechanization is designed to "cut through" the many extraneous and diverse procedures which have been built up over many years and knitting new procedures into a cohesive and coherent entity, using the most modern techniques.

3. Mechanization is the result of scientific programs which convey, through mathematical formulas and the interrelationship of variables, new information which cannot be obtained in any other manner than high-speed calculations.

PART NUMBER SYSTEMS

3

This chapter should be dedicated to those people who have an uneasy feeling their super colossal "hummer" of a "management information system" is running on less than full power. For far too long numbering systems generally, and part number systems specifically have been designed and implemented with very little long-range planning. The impact of a poor part number system has been only vaguely understood. The examples of numbering systems used in this section are not meant to discourage or confuse the reader. It should be remembered that several part number systems explained here were developed in the 1800s, long before any modern management concepts were designed. Not only are these examples used here in an attempt to prevent future horror stories like these for new companies, but also to prepare information which may serve to help change present part numbering systems in other companies, if they are similar to those described here.

Present day manufacturing techniques, when improved by a mechanized system, inevitably create the need for, and the power of, a bill of material system. Such a system becomes a major communication technique and device to convey piece part data and to show the assembly and structural relationship of a product. The changes of configuration due to design alterations and their "effectivity" are reflected in this system. Downstream from engineering, production scheduling, inventory requirements, and purchasing requirements are determined directly by matching a forecast for the product to its corresponding bill of material.

Unfortunately, one of the major deterrents to using this technique adequately in manufacturing is an inadequate, obsolete, and sometimes dangerous part number system.

Part number systems come in an almost endless array. Only the imagination of their designers has limited the varieties and variations. The need to modernize manufacturing systems has caused this problem to become acute. Many part number systems were inadequate at their conception; their dangers were enhanced by the recipients who developed alternate methods to alleviate the problems they encountered. The more modern manufacturing data processing systems which can assist in obtaining large economies by using an integrated data processing system have brought this problem to a head.

It is our intent in this section to describe various part numbering systems which currently exist in industry, their dangers, both physical and philosophical, and finally, to recommend ways to alleviate these weaknesses in manufacturing.

The following numbering systems will be discussed:

Order number oriented

Mark numbers

Group-item numbers

Number matrix

"Standard" parts

Product oriented

Tabulated

Check digits

Significant

Revision level nos.

Drawing numbers

Random numbers

Present numbering systems fall generally into two descriptive segments: (1) significant and (2) nonsignificant. On the surface, this seems to be a simple breakdown. However, when dividing these categories into smaller segments, a bewildering array of number types appears. For example, the part number of one company with a "nonsignificant" system looks like this:

A-XX-XXX-XXX-X

Manually used, it seems to pose no apparent problems, except length. However, due to the variations surrounding the system, the mechanized part numbers look as shown in Figure 3, where "X" means numerical and "A" means alpha. Several comments can be made regarding the metamorphosis from a "simple" (but long) part number to 38 different systems encompassed within it:

1. Of the 6 original segments of the number, the size of all segments were added to or deleted from.

2. Of the 6 involved segments of the number, alpha characters were added to numerical digits, or numerical digits were added to alpha characters, or entire segments were eliminated.

3. Additional significance was added to one segment.

4. The additions and deletions, etc., changing the segment size created the need to "right-justify" or "left-justify" each alpha numerical segment to distinguish and/or "sort" one system from the other.

5. The original "simple" 12-digit part number now requires 19 digits of data to correctly distinguish and identify a specific part number.

Drwg. Size	Pre-Fix	File no.	Signf. Code	Sheet No.	Suffix
A	XX	XXXX		XXXX	X
A		XXXX		XXXX	
A		XXXXX			
A		XXXX			
A		XXXA		XXXX	
A		XXXX	AA	XXXX	
A		XXXX	A	XXX	
A	XX-A	XXX			
A	XX-A	XXX	XX		
A		XXXX	A	XXX	
A		XXXX		XXXXX	
A		XXXX		XXXX w/sign	
A		XXXX		XXXXX	
A		XXXX	A	XXXX	
A	AA	XXXX			
A		XXXX	A	XX	
A		XXXA	XXX	XX	
A	AAAA	XXX	A	XXX	A
A	XX	XXXX	A	XXX	
A	XXXA			XX	
A	XX	XXXX		XXXX	
A		XXXX	XX	XXXX	

Figure 3 "Standardization" of a part number system means not only the length, but the entire format.

6. None of the systems when considered individually and within a manual system exceeds 12 digits.

The importance, or lack of importance, of the significant description built into the system will be discussed later. This example poses the first of several problems which have occurred in the past which should be examined.

When writing a part number in a manual system, such as 10-1, it is not necessary to "fill in" the unused digits which are zeros. When using a mechanized system, however, it is almost mandatory in order that it be recorded correctly for a data processing system. The actual number may not be 10-1, but actually 00010-0001, a not unusual example. Diehards who have used the manual system for years consider this extraordinary requirement ridiculous, time consuming, and worst of all, an excuse to prove the system is inefficient, and uneconomical. They are indeed correct. They are not wrong, the modernized system isn't wrong.

The Part Number System is Wrong

If you require a more powerful method to make your manufacturing system perform, you must design and use tools to make the most of it. We would all think it rather ridiculous to dig the footings for the Empire State Building with a single-hand shovel; why should we hinder the use of a powerful communication system in the same way?

There is a converse problem in this example of using an extremely long, nonsignificant number such as 1160755230A. Long ago in this particular company, employees in fabrication and machine shop areas abbreviated it to a 5230A, or 30A, and "black books" also carried a small description to differentiate it from 1260755230A. Enterprising manufacturing people have been also known to develop their own separate numbering system. An entire company wakes up one morning to find that, to their surprise, the expensive, brand new integrated computer system they just designed and implemented has not increased communication speed one iota, and people everywhere are muttering to themselves about the terrible nonhuman brain—the computer.

Let us take the various numbering systems we listed, show how they are used, and some of the problems they create. Each example is taken from a real and existing manufacturing concern. Only their identities have been hidden. Figure 4 shows the list we will discuss.

ORDER NUMBER ORIENTED: COMPANY ONE

When order numbers are used in the documentation of a design or product, several problems occur, not only in the significant system used, but in the proliferation of additional numbering systems which this method often creates.

In most systems of this type, the order number was originally used as a device for collecting costs. Its effect was to create a "summarized" bill of material which defined all parts and assemblies required for one specific order. Printed for distribution it could look like Figure 5.

While, in this example, parts and assemblies are mixed together because of the order number sequence, in most cases, assemblies are segregated from piece parts. However, the alternate method usually eliminates the listing of parts other than assembly drawings since it is also usual to find no separate drawings for the piece parts. Thus it becomes necessary to consult at least two documents to determine the complete set of components for a product.

The order part number system was supposedly designed to collect costs, but not all costs. Therefore, other numbering systems were designed to collect them as follows:

1. XXXX   Quotation number to collect presales cost
2. XXXXX   Project number to collect presales design costs
3. XXXX   Calculation number for computer calculations

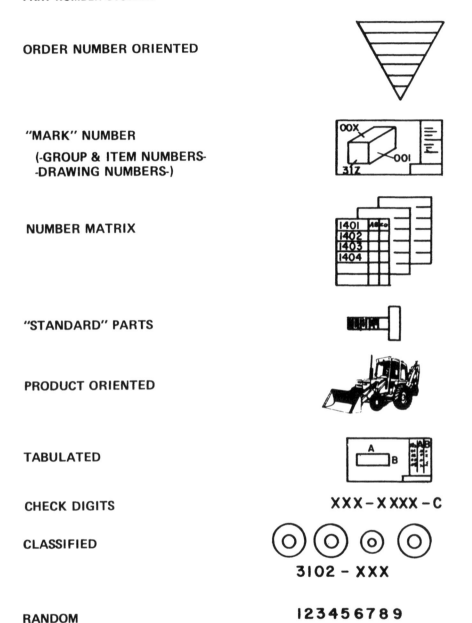

ORDER NUMBER ORIENTED

"MARK" NUMBER
(-GROUP & ITEM NUMBERS-
-DRAWING NUMBERS-)

NUMBER MATRIX

"STANDARD" PARTS

PRODUCT ORIENTED

TABULATED

CHECK DIGITS

CLASSIFIED

RANDOM

Figure 4 The philosophy of part number systems can be divided into nine distinct categories.

Order Number	Item no.	Part no.	Qty	Desc
72-611455	001	1234334	2	Gear
72-611455	002	6455344	1	Bracket
72-611455	003	3437765	4	Motor
72-611455	004	4453437	1	Gear assy
		etc		

Figure 5   Using an order number as a part number immediately nides part which are identical. Thus a system has been devised to "prove" repetition of design does not occur.

As implied before, the order number is normally used to collect costs. However, some other number must be used to distinguish each part and assembly from one another. When this additional number is designated, both the "part number" and the order number must be carried in all documentation — one for cost, the other to identify the parts.

Additional numbering systems existed in this company:

4.   X-XXXXXX   Parts list number a system designed to number assemblies only

5.   XXXXX   Drawing number "old"

6.   XXXXXXX   Drawing number "new" for all "new" designs

7.   AXXAXA   "MPT" part number used to document designs made by a computer

8.   XXXXXXX + XX × XX   "Open" dimension number drawing number plus two dimensions from the part

9.   XXX-XXXXX-X   Machine shop number replaced any engineering part number

10.   XXXXXX   Spare parts number developed by parts department

11.   000000   Hardware part number: it seemed that nobody cared what bolt or nut was used, or that people liked to write data over and over; besides, the parts are cheap, why account for them; store them by description

12.   XX-XXXXXX + XXXXXX   Raw material number includes order number plus part number for specific material

These numbering systems contributed to an acute shortage of raw materials in inventory, because the identity of material by order number made the actual type of material a secondary condition. It was almost impossible to transfer a material from one order to another since the system was sensitive to a receipt of material by order number, not type. Even had a transfer become mandatory, it was necessary to search each separate sales order to find if one of them had identical material, if it had been received, and then make numerous paper work changes to transfer it.

13.   XXXXXX   "Uni" number, a "new" spare parts number to "unify" old systems

14.   XX-XXXXX-X   With the addition of two purchase order systems and one work order system

15. Different numbering systems were devised to collect costs and distinguish parts
     and assemblies

The order number system creates a major problem by generating what could be called
an "inverse hierarchy." Figure 6 depicts this relationship in triangular form. One of
the major problems which results is that information can be passed down from one
level to another (at least when the order number is used). However, it is impossible
or very difficult to pass information up to each successive level without going back
to the master order listing and working back down again. In addition, a minimum
of 14 to a maximum of 17 digits of information must be maintained at all times to
completely identify a part. The addition of a purchase order or work order number
increases this to 25 digits. Later we shall discuss lengths of numbers and the corresponding accuracy of transmission.

Figure 7 portrays a logical or "consistent" hierarchy which not only reduces the
number of systems required from 17 to 7 (even with the addition of 2 new systems),
but also utilizes the full power of the hierarchical or "structural" relationships of a
bill of material system. Using this technique, information will pass up or down from

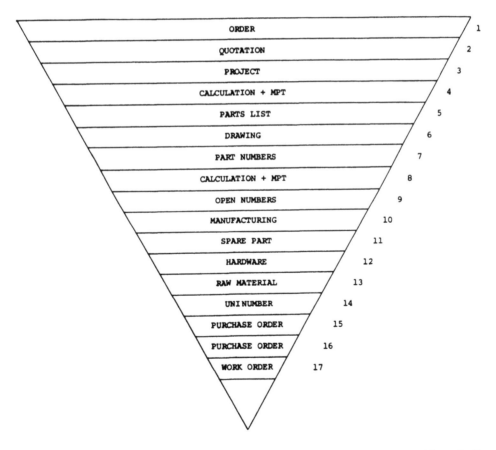

Figure 6 An order number system develops what can be described as an "inverse hierarchy,"
such that the least significant data becomes the most important.

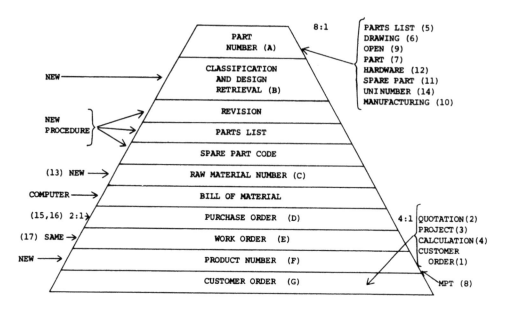

Figure 7 A "consistent" hierarchy builds from detail to logical entities. This hierarchy reflects the real world of data generation and control.

one level to another with equal facility, and eliminate the need to carry an order number along to collect costs or artificially segregate parts and production runs.

It should not be surprising to find several problems resulting from this series of numbering systems which are bound to create disjointed communications. Indeed problems do exist:

1. With all the numbering systems devised, no inventory location system had been developed. It is very difficult to find "standard" parts from memory when no part numbers have been assigned to them, particularly in a 4-floor warehouse storing 10,000 parts.

2. The same data in many different forms and, of course, with different methods for retrieval, is being posted in many different departments.

3. Because it is almost impossible to find old designs, a vast host of design alternatives to perform identical functions are being created.

4. Since rules of interchangeability are not being followed, parts can be changed without changing the part number.

5. Segregation of raw material inventory by order number makes it difficult, if not impossible, to transfer material received for one customer order, but needed sooner on another.

6. Since the system was designed believing all parts would be unique, or totally "job shop," the system documentation perpetuates this theory so that all parts will seem to be unique. (A more specific example will be shown under the Mark Number example.)

7. It is impossible to mechanize the bill of material without a part number for each part and a revision level system.

8. The "open dimension" part numbering system creates many different parts with the same "part" number. It also makes it almost impossible to determine what standard part number format should be used in any mechanized system.

9. Confusion exists between part and drawing number.

10. The system is "hiding" the problem of too many design alternatives by finding a means to record them.

11. Last, but not least, every error of documentation exists:

    a. Same parts have different numbers

    b. Different parts have same numbers

    c. Some parts have no numbers

MARK NUMBER SYSTEMS: COMPANY TWO

I'm sure the gentleman who invented the mark numbering system got a raise. After all, he did save hours of paper work in engineering. No one will ever know, however, how many potentially successful companies were sunk and went bankrupt using this system which creates such an inadequate and expensive means to communicate design data.

No one will ever know how many present job shops could be "mass producing" their products today or selling them 50% less expensively. A mark number system is poison to an efficient, economical, and rapid manufacturing communication system.

Mark numbers were probably invented when sepia reproduction techniques were invented. A sepia is a dingy brown piece of paper which can be used like a master drawing, since it is thin enough to reproduce its own blueprints when put through a reproducing machine. At one time, master drawings were made on linen cloth with india ink, and were very costly and time consuming to make. The sepia set the engineer free to make major or minor variations to a design without making a completely new master drawing.

When an order or a design change was prepared, it was simple to take the master drawing, make a cheap sepia, and then, with special liquid, erase those portions of the design to be changed. New dimensions or design features could then be added to the sepia, a new drawing number assigned to it, and voila, a new design could be sent to manufacturing in just a few minutes or hours. The practice was seemingly innocent, and speed of creating new designs was increased.

Figure 8 illustrates a typical drawing, 12-67712-01. The drawing contains a sketch of the assembly and also a list of materials to be used on the body of the drawing. Figure 9 shows a sepia made from the original, but with design variations, and re-numbered to drawing 12-67515-01. The results of this practice are as follows:

1. A part number consists of the drawing number plus the "item" or "mark" number, or 12-67515-01-005, for example. The finished assembly is given an arbitrary mark number. In this example, the "part number" for the assembly is 12-67515-01-500.

Item	Qty	Description
001	1	
002	1	
003	1	
004	1	
005	1	
006	1	

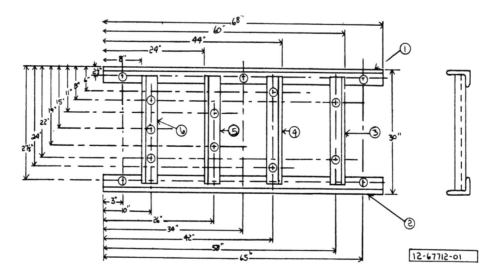

Figure 8 A multipart drawing prevents a designer from reusing parts in another design when they are identical.

2. Figure 8 shows an assembly with six piece parts, each of which are identified by using a specific drawing number and mark number. Figure 9 shows an identical assembly, except for two of the piece parts. Four out of the six parts composing the assemblies are identical. This fact is hidden because they are identified by using the combination of the same mark number but different drawing numbers.

3. Should a conscientious draftsman choose to use any of these parts in his next design, it would be an impossible task, since calling for any of them on a new assembly will require issuing the entire assembly drawing, marking out those parts not needed, and intermixing one drawing mark number combination with a different set of drawing mark numbers on a new drawing. Carried to extremes, a new assembly drawing consisting of 20 parts could require issuing 20 different assembly drawings to collect the required configuration. One European company actually takes such a drawing and cuts it up with scissors to send each part to the proper department for manufacturing.

4. Engineers do not seem to like simple mark numbers as were used in the illustrations. They usually like numbers such as "A1," "K001," or even "S1-½." Can you visualize using a part number like 191200116SL.25? Unfortunately, this is an actual example taken from an American manufacturing firm's drawing.

5. A study was made by one manufacturing firm on the amount of drafting time required to make mark number-type drawings as opposed to single-piece part drawings, a separate assembly drawing showing assembly dimensions only, and a "detached" parts list. In all, 36 similar designs, each composed of 30 piece parts were analyzed. Each original drawing required 20 hours of drafting time for a total of 720 hours. Drawing individual piece parts required only 22 piece part drawings for the entire 36 designs (A size rather than former C size), requiring only 30 hours. In addition, each part within the old system had had shop routings prepared for them. In one example, 21 routings could be replaced with one. Further research into the designs studied showed five different raw material sizes used which could be reduced to one.

Figure 10 shows another kind of drawing which can be made using the mark number system, but with a slight variation. The drawing shows miscellaneous parts which do not necessarily go together as one assembly. This example is also a real one in which the drafter found he had forgotten some miscellaneous parts in several assemblies, and put them all together as shown here. The company used an order number system and issued "shop order pages" to manufacturing. It was a simple matter for the drafter to update these pages to show the additional parts. No one will know how long

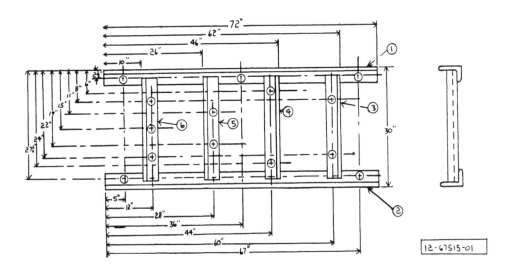

Figure 9 Although only one part has been changed dimensionally, the multipart drawing number changes and therefore, changes all part numbers, even of identical parts.

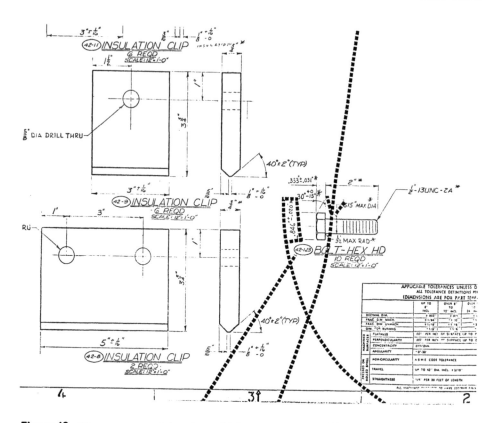

Figure 10 Unnecessary part drawings which are not connected into one assembly create confusion in a manufacturing environment.

it took others to find them. Even if we assume this is a legitimate assembly, the combination of parts on the drawing cause difficulties in manufacturing.

1. The lockwasher and bolt are purchased and are "standard parts" which are, no doubt, in inventory. The two pieces with holes in them are steel and must be routed through the shop for drilling or punching after being routed to the fabrication shop to shear or burn from raw material. The question is, should you send a tote box to inventory control to collect the bolt and nut before or after manufacturing the parts? Also, how long do you think the bolt and nut will last in the tote box before someone picks them up and uses them?

2. The bolt and nut are "standard parts" in that they have been specified by Society of American Engineers (SAE) standards. It is therefore redundant and expensive to make drawings for them at all. In this example, however, the drafter placed a dimension on the bolt which is 0.030 in. different from the SAE standard. Did he make a mistake or did he really wish to remachine the "standard" bolt?

3. Possibly you have been thinking that these standard bolts and nuts could be found in bins on the shop floor as standard items, and would not need to be picked up at an inventory crib. However, a look at the shop order page on which these parts have been placed shows that they require a special chrome-nickel alloy which is not standard. Picture the confusion and uncertainty manufacturing personnel must en-

counter when so many reference documents are needed to make or purchase parts. (A foreman in this same company admitted he frequently took many drawings home at night hoping he could interpret them for his people next day when work was performed using them.)

Figure 11 shows a drawing which is the natural mutation of a mark numbering system, namely multiple assemblies on one drawing. In this example, all the mark numbers are not used on any one final assembly. Parts are added or eliminated according to the assembly required. A table on the drawing shows which parts are required on each different assembly mark number. It takes only a little imagination to realize the errors which can be generated if an analyzer goes down the wrong column, or shifts from column to column, especially on Monday morning. In addition, each assembly requires special notes, so on some old, well marked up drawings, there may be as many notes as there are assemblies, written anywhere, including along the border of the drawing and even upside down. The assembler had best read every note before proceeding.

In order to distinguish one assembly from another on a bill of material, it is necessary to code each column of items. For example, the first column is assembly XXXXXXX501, the second column is XXXXXXX502, etc. Of course, if there are more than 500 parts in an assembly, this practice breaks down, but alpha characters can always be added (and create further confusion).

ITEM	MK # 501	MK # 502	MK # 503	MK # 504	DESCRIPTION
001	1	1	1	1	
002	1	1	1	0	
003	1	1	1	0	
004	1	0	0	1	
005	1	0	1	1	
006	1	1	1	1	

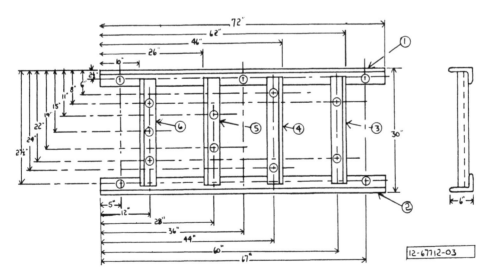

Figure 11 This tabulated assembly creates confusion and potential interpretation problems. Mechanization of the data is complex and often unreliable.

Last but not least, it is difficult to make engineering changes or revision-level changes. Revision changes are usually ignored in this type of drawing system and, since the rules of interchangeability are not followed, it is necessary to carry the revision level along with the part number in order to differentiate one part from another. Some revision changes do not make it necessary to change old parts to the new configuration; some revision changes require that all old designs as well as new designs be replaced.

Figure 12 shows a combination of all the possible variations of mark number confusion:

1. Although the drawing number does not show in the illustration, each mark number, such as 2U55, requires it to make a complete part number. Since there are five variations to the same part on the drawing, only a parts list can designate which one should be manufactured. All of them are not required.

2. Each alpha portion of the mark number is significant. For example, "BP" translate into base plate.

3. Each mark number for the U bolt is also a tabulated drawing, so when it is finally determined which one to make, it is also necessary to pick off dimensions from the table.

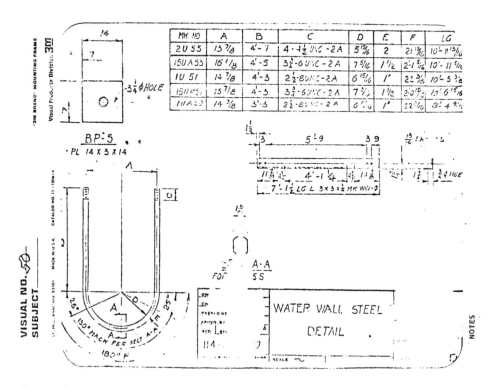

Figure 12 A multipart drawing combined with a tabulated drawing and special notes which alter the pictured design become nightmares to the fabricator. In addition, the final design requires reference to other documents to complete.

4. If, after determining all the data necessary to make the proper part, you begin to manufacture mark number 2U55, watch your step. A note on the drawing shows a round part whose sides have been flattened for that particular design.

An example of the effect this type of documentation has on manufacturing efficiency may be illustrated by the complex and costly results. The example is repeated word for word taken from a memo written by the production control manager.

The following statistics apply only to data and paperwork requirements to accommodate the revised design configuration. The requirements for the original design are not included. An average of three revisions was experienced in converting from the original design to the latest revision.

Requirements if the revisions had been made in one step:

Activity involved	To make parts	To make assy's	Total
Mark numbers involved	44	26	70
Pages of processing	367	3320	3687
Process descriptions	44	26	70
Process routings	816	12700	13516

Requirements as the changes occurred:

Activity involved	To make parts	To make assy's	Total
Mark numbers involved	44	26	70
Pages of processing	779	10117	10896[a]
Process descriptions	44	456	500
Process routings	3792	37740	41932

[a]This was the actual number of pages of shop processing distributed to the shops as a result of the design changes.

It is not possible to ignore these problems if a company is embarking upon a program for mechanizing their bills of material or CAD/CAM. Some companies have tried. They have been forced to change their procedures after many months and dollars of effort.

GROUP ITEM NUMBER: COMPANY THREE

An extreme example of this type of system occurred in a manufacturing company where each assembly and part are given a "pseudo" number which classifies the product into groups and levels of parts within the product. Figure 13 shows an example of this renumbering.

Cost for each assembly is collected by ignoring part numbers and sorting all costs

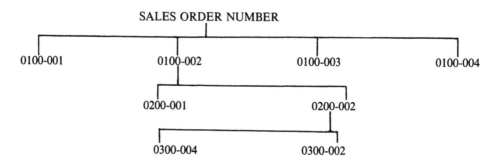

Figure 13 This implementation of a group and item number essentially renumbers all parts and assemblies documented in engineering. The result is confusion, lack of communication, and an artificial job shop.

by "group." This semiclassification of parts creates an additional complexity in the system in that each new order inevitably has a new set of sequences for group and item numbers. It takes a combination of order number and group and item numbers to identify a specific part or assembly. The strange, but actual result is the creation of a firm belief within the company that nothing identical is ever manufactured and that their operation is 90% job shop. The exceptions are considered to be "standard" parts, whose use is so obvious that a separate numbering system is attached to them to remove their manufacture or purchase from the strangulation of the group-item number procedure. With an 8-digit order number, an 8-digit group-item number, and a 15-digit part number, it requires 32 digits of information to positively identify and locate a specific design. This does not complete the numbering system, since an additional work order number is also utilized to keep track of labor performed.

It is possible the reader by now may believe these are extreme examples, and, indeed, they may be. Nevertheless, other systems cause many of the same communication and transfer of data problems. It seems evident that an engineering change, for example, on a particular customer order requires several translations to fully implement it. It is comparable to a meeting of the United Nations where the speaker must have his words translated into many different languages as he speaks. Unfortunately, manufacturing companies are not this efficient. The result, then, is a continuous series of translations between the engineering part number system, the job order number, and the group and item number. Add to this the rules of interchangeability and revision-level changes to designs and it becomes apparent that the mechanization of the system is difficult or impossible and errors will be high.

NUMBER MATRIX: COMPANY FOUR

The numbering system described here has no official name, so the title has been created as a description of the process. Figure 14 shows a series of master cards (in the actual example, these were well dog-earred green cardboard). Some chief engineer spent 10 minutes one day to create this system, and watched the octopus grow over the next 50 years.

The system starts simply enough. When a drafter prepares a first drawing, a part number is selected using one of the master cards. The number chosen may be from

1400	A	B	C	D	E	F	G	H	I
1401	√	√							
1402	√								
1403									
1404									
1405									
1406									
1407									
1408									
1409									

(Overlapping sheets behind: 1410 A B C D E F G H I; 1420 A B C D E F G H I)

Figure 14 Use of a part number system which generates random part numbers, not only in designators, but in length, eliminates accuracy, consistency, and retrieval.

row 1401, column A, for example. By placing a check in the appropriate square, drawing number 1401A has been selected, and no one else can now use the number (theoretically, as you will see). Another drafter, upon completion of a drawing may select 1402A. When the first drafter designs another part or assembly similar to the first design, it is only "natural" to want to file it for future reference near the first design. However, 1401B does not seem to be too close, or has already been used; 1402A has also been taken. Therefore, the drafter selects drawing number 1401A1. Still later, two more similar drawings are made and may then be numbered 1401A1A and 1401A1B. The drafter now has a choice of starting over with another personal numbering system, or continuing in this square. By now, it is also too late to issue a drawing 1401B, since the drawing will not be filed near a similar assembly drawing made previously. Consequently, over five or ten years, the drafter will be issuing a drawing number such as 1401A1A21B. Please remember, an "item" or "mark" number must be added to the drawing number to create a part number.

In this particular company, a machinist approached one of the machine shop foremen with a perplexed look. In his hand was a drawing of a steel shaft 4 in. long and ½ in. in diameter. On the drawing was a comment, "stamp the part number onto the end of the shaft."

"STANDARD" PARTS: COMPANY FIVE

Standard parts are usually found in an engineering standards manual. Since standard parts usually mean "hardware" such as nuts and bolts, looking up the proper part and its corresponding part number is a daily annoyance. The really valuable "standard" parts such as castings, forgings, and fabricated parts designed by the company are usually never found in the standards manual. Thus, the original intent of such a designation—using the same part many times—is not achieved for the more important items in a manufacturing company.

It may be suspected that standard parts originated from drafting manuals which originally showed how to draw purchased items. The natural "add on" to this set of instructions were part numbers and dimensions. Unfortunately, everything in an engineering standards manual seems stagnant and "out of date." The means for changing from one set of standard parts to a new set which might reflect the dynamics of the changes to the manufactured product is usually beyond the scope (and comfort) of standards personnel. The following excerpt comes from a letter critiquing the standard parts program of a large earth-moving equipment manufacturing company:

> The original approach taken by _____ company in setting up a standard parts catalog was to place parts having an obviously high usage or miscellaneous parts such as bolts and nuts at an engineer's disposal. The original concept was based on the special needs of the time aimed at keeping such parts (particularly purchased parts) at a minimum. This program deteriorated over the years:
>
> 1. Obsolete parts are not removed from the catalog.
> 2. New categories are rarely added to the list.
> 3. Usage is not necessarily a criterion for entry into the catalog nor for its removal.
> 4. Few, if any, parts manufactured by the _____ company, are in the catalog, thereby eliminating any method for standardizing internal parts manufacture.
> 5. There is no method available for collating similar parts to determine usage or standardization requirements.
> 6. No major program or policy to enforce usage of standard parts has been implemented.
> 7. Parts which have been scrapped due to obsolescence still remain in the catalog.
> 8. Standard parts should be determined by usage, economical studies, and practical considerations rather than by the whims of a standards group which has no solid information upon which to base their decisions.
>
> The rapid rate by which new parts are being released by the engineer department should necessitate a farsighted, vigorous program to collate and classify parts into categories for easy reference and to keep constant watch on superfluous parts. The present standard part program has lost the respect of engineering due to being inflexible, antiquated and limited.

Perhaps there are other comments which could be made about standard parts. Probably the key problem is merely the need to define what a standard part should be. Obviously some sort of retrieval system would assist designers and engineers. We will discuss classification and design retrieval later.

PRODUCT ORIENTED NUMBERS: COMPANY SIX

Many companies have added significance to part numbers by imbedding a product code as a suffix or prefix to the part number. Rarely, however, has a product-oriented part number caused such a severe problem as found in the company from which the following example is taken.

The manufacturing company being discussed began issuing part numbers by product number as shown below. The different prefixes represent different products:

CHB-1

CP-1

X-1

SB-1

The next part designed for each product had the following numbers:

CHB-2

CP-2

X-2

SB-2

Etc.

Soon it was felt necessary to code the product so that competitive information was not so easily discovered. The number system was therefore converted to:

001-00001

008-00001

037-00001

Etc.

Confusion between identical suffixes tied to different prefixes caused the next change. In addition, parts designed for one product were being used on many products and it confused shop personnel. Should they change the part number? A new product code was implemented, but the suffixes were applied serially:

941-5001

962-5002

941-5003

037-5004

941-5005

Etc.

Soon the company became large enough to hire a clerk to ¨le the drawings. It was difficult for the clerk to find where to file them, or look for them. Accommodatingly, the company added a drawing size to the part number:

941-A-5001

962-B-5002

Etc.

In a very short time, the company doubled in size; a computer was hired, and the quantity of drawings also increased rapidly. The size of the part number was on the verge of requiring an increase in size. Tabulated drawings were introduced, so the part number became 941-A-7800-23. In desperation, the company decided to replace the system with an 8-digit, nonsignificant number. After a few traumatic days during which people discovered that the new number system told them absolutely nothing except to identify a specific part, the program was accepted with little comment, and certainly no documentation problems. The system has now been in use for 10 years and the company continues to increase in size.

TABULATED DRAWINGS: COMPANY SEVEN

It appears that every company in the world uses tabulated drawings. They have been used and abused to such a degree, but are so popular, that the slightest criticism of their effectiveness causes indignant responses. Figure 15 shows a tabulated drawing. The drawing number and item number must be used in combination to create a part number. Finding a part number may be easy; determining what the part actually looks like and how to make it is not apparent. The simplicity of the picture is disarming also, even if the number of tabulated dimensions seem formidable.

Figure 16 shows an extraction of some parts from the tabulated drawing.

One part has four, not five holes, not equally spaced.

One part has four, not five holes, not equally spaced, with different diameters.

One part has two, not five holes.

One part has one, not five holes, on a square piece of material.

So much for clear tabulations.

Figure 17 shows another rather messy tabulated drawing. Notice that "T" is used to denote both thickness and diameter and "D" is used both for width and diameter. "D" and "T" are also both used to denote the same dimension on the same part.

There no doubt was a twofold intent in the original use of these drawings, retrieval of similar parts, and elimination of drafting time. Of course, the fact that one must memorize the drawing number or record it in a black book somewhere to find the drawing is ignored. Many engineers have made tabulated drawings for their own personal use, allowing other engineers to make up duplicate series of parts blithely unaware of its existence.

Probably the most amazing misuse of a tabulated drawing occurred in a company which manufactured power transformers. Engineering practices not only included tabulated drawings, but also tabulated assemblies, and order number bills of material. ("Manufactured" is in the past tense because they are now out of business. No longer competitive in manufacturing; could it be because of their documentation methods?)

One engineer found that searching for a very commonly used tabulated drawing was time consuming. It was always on some other person's desk. There were 200 shafts dimensioned on the drawing; everyone knew it existed and everyone used it. One day, frustrated at not being able to find it easily, the engineer made a sepia of the drawing, changed the drawing number and triumphantly proclaimed he now had an easy

33

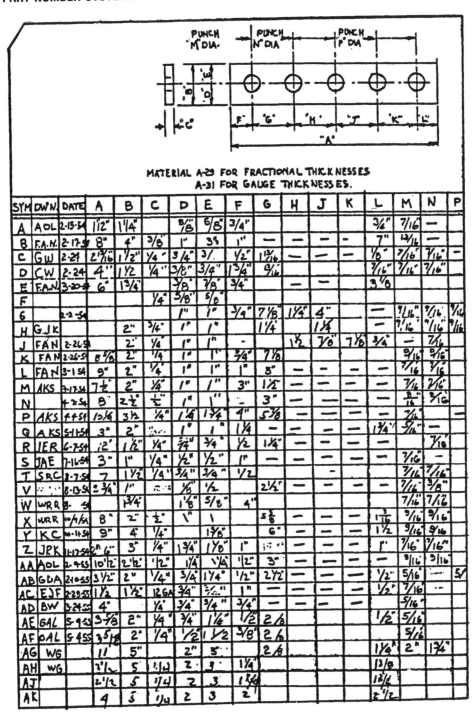

MATERIAL A-29 FOR FRACTIONAL THICKNESSES
A-31 FOR GAUGE THICKNESSES.

SYM	DWN	DATE	A	B	C	D	E	F	G	H	J	K	L	M	N	P
A	AOL	2-13-54	1½"	1¼"		3/8	5/8"	3/4"					3/4"	7/16"	—	
B	F.A.N.	2-17-5	8"	4"	3/8"	1"	3/8	1"	—	—	—	-	7"	11/16	—	
C	GW	2-21	2 7/16"	1/2"	1/4"	3/4"	3/.	1/2"	1 3/16	—	—	—	1/8"	7/16"	7/16"	—
D	CW	2-24	4"	1½	1/4"	3/8"	3/4"	1 3/4"	9/16				7/16"	7/16"	7/16"	
E	FAN	3-20-5	6"	1 3/4"		3/8	7/8	3/4"		—	—	3 1/8				
F					1/4"	5/8"	5/8"									
6		2-2-5				1"	1"	3/4"	7 7/8	1 1/4"	4"		—	7/16"	7/16"	1/4
H	G.J.K			2"	3/4"	1"	1"		1 1/4		1 1/4		—	7/16	7/16	3/16
J	FAN	2-26-5		2"	1/8	1"	1"	-		1/2	7/8	7/8	3/4"	—	7/16	
K	FAN	2-26-5	8 5/8	2"	1/4	1"	1"	3/4"	7/8					9/16"	5/16"	
L	FAN	3-1-54	9"	2"	1/4	1"	1"	1"	5"	—	—	—	—	7/16	7/16	
M	AKS	3-17-54	7 1/2	2"	1/4"	1"	1"	3"	1 1/2	—	—	—	—	7/16	7/16	
N		4-2-54	8"	2 1/2	1/2"	1"	1"	-		3"				—	3/16"	3/16	
P	AKS	4-4-54	13/8	3 1/2	1/4"	1 1/8	1 1/4	4"	5 3/8	—	—	—	—	7/16		
Q	A KS	5-11-54	3"	2"	...	1"	1"	1 1/4	—	—	—	—	1 3/4"	5/16"		
R	IER	6-7-54	12'	1 1/2"	1/4"	3/4"	3/4	1/2	1 1/4"	—	—	—	—	—	7/16	
S	JAE	7-16-54	3"	1"	1/4"	1/2"	1/2"	1"	—	—	—	—	7/16"	—		
T	SRC	8-7-54	7	1 1/2	1/4"	3/4"	3/4"	1/2			—			7/16	7/16	
V		8-13-54	2 3/4"	1"		1/8	1/2		2 1/2"					7/16	3/8	
W	WRR	8- 5		1 3/4"		1 1/8	5/8"	4"						7/16	7/16	
X	WRR	10/4/54	8"	2"	1/2"	1"	1		5 1/8	—	—	—	1 7/16	9/16"	9/16"	
Y	KC	10-14-54	9"	4"	1/4"		1 7/8"		6"	—	—	—	1 1/2	9/16	9/16	
Z	JPK	11-17-54	6"	3"	1/4"	1 3/4"	1 7/8"	1"		—	—	—	1"	7/16	7/16"	
AA	AOL	2-9-55	10 1/2	2 1/2	1/2"	1/4"	1/4	1 1/2"	3"	—	—	—	—	9/16"	9/16	
AB	GDA	2-10-55	3 1/2"	2"	1/4"	3/4	1 1/4	1/2"	2 1/2	—	—	—	1/2"	5/16"	—	5/
AC	EJF	2-23-55	1 1/2	1 1/2"	12GA	3/4"	3/4"	1"	—	—	—	1/2"	7/16	..		
AD	BW	3-24-55	4"		1/4"	3/4"	3/4"	3/4"	—					5/16"		
AE	GAL	5-9-55	3 7/8	2"	1/4"	3/4	1 1/4	1/2	2 1/8				1/2"	5/16"		
AF	OAL	5-4-55	3 5/8	2"	1/4"	1/2	1 1/2	3/8"	2 1/8					5/16"		
AG	WG		11	5"		2"	3"		2 1/8				1 1/4"	2"	1 3/4	
AH	WG		2 1/2	5	1 1/4	2	3	1 1/4					13/8			
AJ			2 1/2	5	1/4	2	3	1 3/8					13/4			
AK			4	5	1/4	2	3	2"					2 1/2			

Figure 15 This tabulated drawing is typical of how Engineers thought they had designed a "short cut" method to eliminate paper work.

TABULATED DRAWINGS

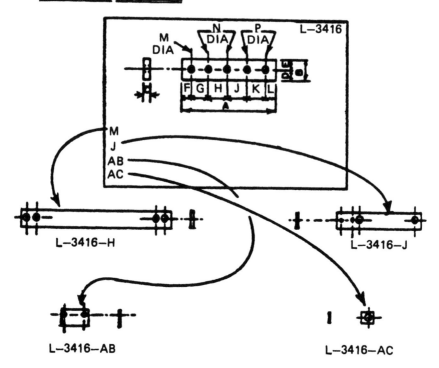

Figure 16 The results of a tabulated drawing and lack of accompanying discipline show up here. A combination of parts other than that pictured on the tabulation results. Thus a tabulation creates more problems than it solves.

method for finding and using the drawing, thereby saving himself much time. The critical result of creating 200 new shaft part numbers made absolutely no impression on him.

If there are some good reasons at all for tabulated drawings, what are some of the reasons for eliminating them or discontinuing their use?

1. Even though one of the original reasons for a tabulated drawing was retrieval of parts, people sometimes tend to be a little lazy. Once 10 to 20 parts are shown on a drawing, we tend to feel it is easier to add a new one than look to see if the one we want is already there. Therefore, we will often find "duplicate entries" on a tabulation.

2. Inevitably "part numbers are wasted," because provision must be made in a part numbering system to accommodate the item number as part of the part number. Since only a small percentage of all numbers become tabulated, the remainder must be utilized carrying zeroes in those positions.

   (As an experiment, think of an 8-digit part number as 100,000,000 numbers. If two of them are reserved for tabulation, 99% of the total numbers available have been eliminated from use. Shocking isn't it?)

3. The more complex drawings are "confusing to the fabricator." Figures 15 and 17 are examples of such confusion. There are so many dimensions to remember,

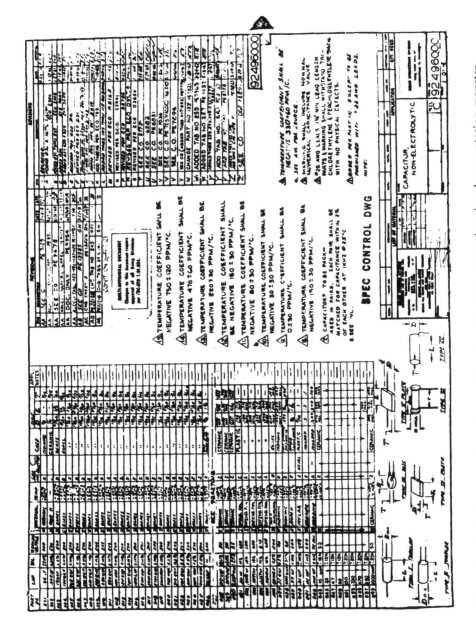

Figure 17 This tabulation shows how no discipline can create an engineering documentation monster.

35

and the final part does not look anything like the drawing; so, no doubt, the fabricator makes his own drawing before making the piece. If it's Monday morning at 7:00, after a long weekend, it is also possible to wander back and forth between columns and make an entirely new design.

4. When a minor change is made to one part on the tabulation, "revision-level control is difficult." It has never yet been heard that only one part on a tabulated drawing may require a change. Instead, it is said with great conviction that any change required will affect all the parts. It seems to the writer this philosophy is simply not true. In companies where parts which could be tabulated are not, it is possible to find various minor changes made to some, but not all, the parts.

5. Very often tabulated drawings "require special handling" through manufacturing. If a tabulated part has several operations to be performed on it, each machine operator must reinterpret the drawing. Tabulated assemblies may require redrawing in manufacturing simply because a variety of totally dissimilar parts are used in various assemblies.

6. The tabulation of 20, 30, or 40 parts does not show any discrimination, and therefore "does not show preferred versus nonpreferred items." The result of this omission is to allow engineers to use any part they wish, even though some of them may be very expensive, obsolete, or simply poorly designed.

7. Many companies do not utilize bills of material, and therefore rely on *order point control* inventory reporting to determine part usage. In such a system, when new parts are released, an educated guess must be made on new usage. Issuing 30 or 40 parts on a tabulation can cause panic, confusion, or worse, complete unconcern. In any case, disaster and *inventory stock-outs* are possible.

8. Some designers have the knack for making a tabulated drawing very tantalizing by apparently dimensioning a complete part. However, it is possible to find tabulated drawings "requiring reference to one or more additional documents."

9. Even if each part on a tabulated drawing is not placed in inventory, some companies prepare routings and other necessary paper work for each part. If they do not, the eventual active release of each additional part will require a search to see if the part has been used before. In any case, the system "creates unnecessary paper work."

10. Tabulated drawings "do not assist in quick design retrieval." As mentioned previously, most people must write the drawing number and perhaps the title in a *black book* to find it. When discussing significant part numbering systems, it will be pointed out how dangerous looking for a part by name can be.

11. "The reference drawing illustration does not always reflect the tabulated dimensions." It is easy to take the picture of a shaft with two diameters and show the dimensions of both as equal. Even more complex drawings can be thus altered and confusion will be the result.

CHECK DIGITS: COMPANY EIGHT

With the advent of keypunches and computers, a heyday was created for systems analysts. It became their responsibility to check whether people wrote down numbers correctly. Perhaps it happened the other way. Keypunches and computers force people to write down a lot of numbers—correctly. The systems analysts were forced to find a method to prove the computer was right and the human was wrong. In any event, many companies added a check digit to their part number and, in some cases, every other numbering system in the company. Perhaps it is wise to explain to the uninitiated what a check digit is and how to create one. First, take a normal 10-digit number, say 12-7998-0750. Second, multiply every other number by 2 starting at the "units" position as:

```
1  2  7  9  9  8  0  7  5  0
   ×2    ×2    ×2    ×2    ×2
```

Third, add the sum of these calculations, or:

$$1 + 4 + 7 + 18 + 9 + 16 + 0 + 14 + 5 + 0 = 47$$

Fourth, subtract the results from 100, or $100 - 47 = 53$.
Fifth, truncate the number, meaning remove the first digit, leaving 3.
Finally, add the number to the part number, or:

12-7998-0750-3

Before fainting, be hastily assured that a computer calculates and prints a list of these numbers when they are needed. (It should also be noted that this is one example of a check-digit calculation. Other algorithms exist.)

There is a specific reason for a check digit. Whenever a part number is keypunched on a special keypunch, the machine keeps track of the digits punched and forecasts, or computes what the eleventh digit should be. If the keypunch operator punches "3," but has made some error previously in punching the number, the keypunch yells "tilt," and locks itself up (computers check direct input in the same way).

Of course the keypunch operator can also make a mistake in punching the check digit, and the same events will take place. This system is credited with finding many errors. We should be concerned with some of the fallacies of this system as well as some of the inadequacies.

1. It is well known that the length of a number increases the probability of error in writing the number down or conveying it verbally to others. Obviously, a check digit makes a part number longer. In fact, a study made in the company under discussion found that 50% of all errors made in writing down part numbers occurred when writing the check digit.

2. If a checking system is to be completely reliable, it should be effective under all conditions. Nevertheless, many documents are handwritten and never go to a computer. The check digit is useless.

3. An interesting paradox occurs in engineering when a check digit is used. Assume an engineer makes a drawing, and places the part number 12-7998-0750-2 on it. Suppose an input form for the computer is properly completed with 12-7998-0750-3. The computer will "okay" the input, while the drawing will probably be used a few times, if not forever, and those who use it will innocently stamp, mark, or handwrite an incorrect part number on many documents.

4. The majority of manufacturing concerns in the world are profitable and successful and do not have a check digit saddled to their number systems.

SIGNIFICANT NUMBERS: COMPANY NINE, AND ALL

The second most emotional problem in a manufacturing company besides designating the length of a part number, if whether it should also have significance. It is apparent that those who held out for significance won; almost every company has incorporated some meaning into their part number system.

If we should list those pieces of information which are classified and included as part of the part number in the companies we have discussed, it might give us a clue to the type of information engineering people, and others, feel is important:

Drawing size	Division of company
Product used on	Drawing number
File number	Drawing sequence
Sheet number	Raw material code
Part dimensions	Part type code
Drawing index location	Make or buy
Mark number	Vendor part number
Assembly identity	Part list number
Revision level	Computerized part
Tabulated drawing number	Commodity
Check digit	

It would appear these 21 pieces of information are important to only a select group of people and do not necessarily add one real dollar to the profit of the company. Furthermore, many of these pieces of data such as make or buy, vendor number, etc., can change, thereby changing the part number. It also looks as though someone again took 10 minutes from a lunch hour to design this data inclusion. After all, there are some really sophisticated numbers that classify products in the suffix and part type in the suffix.

Inevitably, significance fails in a numbering system because amateurs classify. There are more magazine articles proving significance in a part number does not work than there are those promoting the wonders and economics of significance. It is difficult to assess which is most difficult, to remove an inadequate significant numbering system, or implement a good one.

There are all sorts of "systems" installed today. An example of fuzzy thinking appears in the description of a classification system designed and developed by a "national association" through their designated "modern methods committee." It may clear the air to quote from their brochure:

What is a standard numbering system? The _____ standard numbering system is a plan which will bring about a similarity in numbering format from one manufacturer to another the complete number for any item is in two parts. The first is a 4 digit manufacturers code number assigned by _____. The second is a 5 digit item code number assigned by the manufacturer. The complete 9 digit number will identify any product item in the industry. Each participating manufacturer is completely free to assign any 5 digit product numbers to his line, because the 4 digit manufacturer prefix differentiates his number from those of any other manufacturer. . . . Here is an actual number that stands for a hand tap, two more digits are concealed around the curve of the tap. For those who have the code, it tells the complete story about the tool, including size, thread, flutes, style, finish, etc. For order entry, picking, or receiving, however, it's worthless because of its length and complexity. Significant numbers tend to be long and confusing. They are also subject to revisions as products are modified and additional significance added, if required. The best approach to product numbering is a sequential, non-significant number. Significance is not needed by distributors. We believe leading computer suppliers will agree with this statement.

Item numbers under the _____ standard numbering system are assigned by the manufacturer and are completely under his control. There is no need for similar products to carry the same five digit number unless the manufacturers desire this".

Several awesome errors or misconceptions appear in this particular blurb. First, it is important to realize the five-digit number mentioned is to be established by the user. If a 6,7,8, or 9-digit number already exists, the user is in trouble. Five digits is, of course, not the whole system, since the *standard number system,* which is treated as nonsignificant, is in reality a 9-digit system with significance, because the first four digits are assigned one to each company that decides how to use it. If that doesn't make it significant, I'll eat my hat. (Why use five digits anyway? The five-digit length would allow 100,000 companies to be included, and there aren't that many in this particular business. Also see below.)

Believe it or not, this marvelous new "invention" (system) took the modern methods committee two years to research and involved 125 manufacturers' representatives (100 companies are now participating). Some of the misconceptions in their brochure need clarification:

1. A code number should not have to tell a story which everyone can read. Until we can logically define what the significance is for and who uses it and for what purpose, the entire idea is academic.

2. The example, describing the retrieval of many attributes of a tool, describes a *polycode,* or multiple attributes, which are difficult to condense into four or five digits.

3. Stating that the number is useless for order picking, etc., because of its size makes sense for any number system, so why just pick a significant one? Will an additional number be required for this activity?

4. If changes to a part or assembly change the significance of a number, the numbering system is incorrect, or the rules of interchangeability are not being enforced. For example, when a change to a part causes a change to the code number in a good classification system, it always means a new part number must be issued.

5. Relying on computer manufacturers to agree with these principles is a real "cop

out," since the only significance any numbering system has for the computer is the number itself.

6.   It would seem practical to assume that any significance should be used to catalog similar items for easy reference. Not only does this article say it is unimportant, the numbering system they are selling is designed to signify differences only, not similarities.

A major automobile manufacturer uses a significant numbering system where each digit or set of digits is used to designate one characteristic and its variables. The system is called a *polycode.*

Their numbering system is as follows:

Class	B	
Item	A	
Type of item	A	
Standards	X	
No. of drawings	2	
No. of sheets	2	
Diameter		A
Length		B
Type of finish		2

or

BAAX2-2AB-2

Of these 9 codes, only 4 describe any of the physical attributes of the part and, of course, this is limited to 9 characteristics each. It is this apparent unwieldiness of a coding system that makes some nonbelievers in significant systems. It seems obvious also that the number, with its mixed alpha numerics will take a genius to transmit or remember it.

Let us suppose we decide to become experts at classification and develop our own system. If we are smart we should start with something simple such as materials, and using definitive nouns, classify them. We will possibly end up with a system similar to one below:

1000 Materials
  1100 Production materials
    1111 Brass
    1112 Sheets
    1113 Bars
    1114 Castings
    1115 Tubing
  1120 Steel
    1121 Plates

1122 Strips

1123 Wire

1124 Bars

We seem to be home free, and have loads of room for expansion. We then apply the code to our materials. In our theoretical company, we find 700 steel plates, 4 wire sizes, 7507 strip sizes, and no bar. We have discovered another problem. Certainly the code is all-embracing, with logical descriptors. We have simply forgotten to check frequency or volume of use. To fully utilize a classification, requires special tailoring. No red-blooded engineer likes to skim through more than 15 or 20 drawings or sizes to find what is wanted. Therefore, better rules must apply. Later, when we discuss classification principles, we will discover some of these rules.

REVISION LEVEL NUMBERS: COMPANY TEN

Revision-level numbers are usually thought of in a different context from part numbers. Unfortunately, lack of good part number "change" rules forces inclusion of the revision level as part of the part number. Therefore, it forces a major alteration in the company's actions to understand the reasons for and the results of its inclusion. The company used in this example has a particularly difficult problem since it has made a wide variety of electronic products for many years, and the lack of recognition of the problem and its implications will take considerable change and education to adapt.

When products are designed, they are documented using part numbers and shown in combination using a bill of material or parts list. Throughout the life of a product, many engineering changes are made, and it is therefore necessary to devise a means of recording, identifying, and communicating these changes.

It is important to define what a revision level is and what it does:

1. It is not a part of the part number.

2. Parts cannot be inventoried under various revision levels. (They are interchangeable.)

3. A revision level is a "signal" to notify and communicate to other people that a change has occurred.

4. It is a means to record many sequential changes which can occur to the same part number to "keep track" of the stages of change.

5. It should be used as the "signal" on the following documents:

    a. Purchase orders

    b. Work orders

    c. Routings

    d. Parts lists (the assembly number only)

    e. Engineering change notices

These rules are designed to give confidence to people who use drawings, parts lists, and work orders that a part or assembly has a positive, one-of-a-kind identity.

The company under discussion did not follow the rules cited above. Changes made to a part or assembly did, however, require a change in revision level. A numerical revision level change, as from 01 to 02 implied a noninterchangeable change. An alphanumeric change such as 01 to 02A implied an interchangeable change. Unfortunately, manufacturing and inventory control is managed without this added number appearing on all documentation. Since the differences between a numerical change and an alpha change was not well understood, many problems occurred. For example, Figure 18 shows the history of 10 changes to one design. Without strict rules, it can be seen that the ten "revision" changes include only 3 required part number changes. Since this difference has not been well defined, the addition of the revision level to make it part of the part number only increased the amount of documentation and confusion. Not only will part numbers change more rapidly under these conditions, but also identical parts will be hidden by the different part numbers.

In addition, this company has many documents which are required to completely describe a part or assembly. Each type of document has its own revision-level sequence. It is therefore difficult to track which document change causes a part change. It is also impossible to determine the part number of the final part, since a schematic diagram, with the revision level as part of the part number may have part number 2211-4465-04, while the actual drawing for the part may have part number 2211-4465-16. A general mixture of drawings or specifications which may affect a piece part only, the assembly the piece part is used on, or both, creates the ultimate confusion. It should be added that many designs have several sheets. Each sheet of the design could have different revision levels depending upon where and when a design change occurred.

The misuse of a revision-level system has serious and far-reaching implications. Each spare part order must go through some sort of interpretation. Vendors who man-

Original part number 1234-3010

Changed part no	Engineering change	Correct number
1234-3010-1	Document chg	1234-3010
1234-3010-2	Added holes	1234-3011
1234-3010-3	Document chg	1234-3011
1234-3010-4	Hole change	1234-3012
1234-3010-5	Document chg	1234-3012
1234-3010-6	Document chg	1234-3012
1234-3010-7	Delete holes	1234-3013
1234-3010-8	Document chg	1234-3013
1234-3010-9	Document chg	1234-3013
1234-3010-10	Document chg	1234-3013

Figure 18 When the revision level is used as a part of the part number changing it cannot designate whether the change is interchangeable or not. Thus a proliferation of new parts which are really identical occurs.

ufacture parts must be kept constantly vigilant to be certain all changes which are real changes are incorporated into the finished design.

In another company, a different and quite serious problem occurred through the use of part numbers. The title of the form was the culprit in this example. Normally, each part number for a part or assembly was changed according to the rules of interchangeability. However, the "top-level" part number for the product is a separate system, with a different format, called the *master list*. The number format is as follows:

Product type   XXX

Identification no.   XXX

Sequence number   XXX

Revision level   XX

Changes made to the product bills of material caused many revision-level changes to these master list part numbers. When such a change occurred, documentation was updated and manufacturing proceeded. Suddenly one day it was discovered that current orders in the shop included the same master lists with revision level 01, 03, and 05. This was an obvious error, since revision level 01 and 03 had long been obsolete and discarded. An investigation discovered the terrible fact that what was thought to be a revision-level change was in fact a part number change. All configurations being manufactured were using current master lists for valid products. The form upon which this data was transmitted had been wrong for 10 years. It required some diligent and expensive redocumentation to reconstruct all the proper data.

DRAWING NUMBERS: ALL COMPANIES

In describing various number systems, it has been implied that a drawing number is different from a part number. This is not always true. Whenever one piece part or one assembly appears on a drawing and those items are identified with the same number as the piece of paper called a drawing, the two are identical.

In modern data processing systems as well as in configuration control techniques, this one-to-one relationship makes a critical and positive contribution to the economy of documentation and identification. The use of drawings whose part numbers do not identify discrete entities creates several bad practices or problems:

1. If mark numbers or item numbers on drawings are standardized to show identical parts, such identity would be obscured by adding the drawing number to them.
2. If all identical parts were set up with the same mark number, it would require a larger (6–8 digit) mark number, and corresponding update and alteration to all drawings.
3. Since interchangeability rules are usually not followed, a revision-level change on a drawing change is significant for each piece part on the drawing. Conversely, if a change is made on one piece only, the mark number as well as the revision level of the drawing would also have to change.

4.   It is not often clear whether the "same" part and "identical" part have the same meaning.

5.   In order to identify revision-level changes on a piece part appearing on a drawing, a mark number revision-level change system must be implemented. However, the drawing number revision level cannot then be carried as part of the part number. If it is, two revision levels must be maintained for each part.

6.   Since different types of parts appearing on the same drawing require different routings, all parts must be routed through unnecessary work stations and must be completed in a "bunch" or "lump," or a series of copies of the drawing must be issued for each mark number with the part to be made circled.

RANDOM NUMBERS: HOPEFULLY

What should be used as a part number? In summary, it should be used as a key to unlock the door of a data base to obtain all the other data needed. An empirical set of rules may be constructed to define a part number:

Nonsignificant

Numerical only

Sequentially issued

Eight digits maximum

Divided by dashes as XXXX–XXXX

Numbering system problems won't just go away unless a company goes out of business. Recognition of the problem must be first, of course. Although several companies have been described here, hundreds, perhaps thousands, of companies are using some variation of the number systems described here. Modern methods for changing them are available. The economies to be found in modernizing numbering systems for better communication of data is sizeable. Let us look at some of the potential solutions.

HOW CAN A COMPANY CHANGE?

Let us suppose you are employed in one of the companies we have described. What can be done to change the system? Should it be changed? What is the first step?

If a recommendation is made to go back and change every existing part number, the person who makes such a recommendation will be ignored, or fired, or the company may go out of business trying to effect the change. Any such major change is, for a large company, equal to changing all the names or all the addresses of everyone in the city of Chicago. It requires careful planning. Here are some suggestions for improving and changing:

1.   While it is not logical or economical to change all old part numbers, it is possible to change all new ones. Establish a cut-off date for the conversion. Recognize that the old system or systems will be around for some time and allow for this.

2.   Recognize that data is hierarchical in a company. do not allow extraneous,

or even important data to become incorporated into the part number. The part number is meant for only one use—as a key shown in Figure 19.

3. Realize that the length of the part number creates an inverse accuracy for transmitting it. Several studies have been made concerning lengths of numbers and corresponding accuracy. In one scientific study, several different kinds and sizes of numbers were recorded on tape. Many persons were asked to listen to these tapes and then pass what they heard on to other people, or write it down. The summary of accuracy is as follows:

Number size	Accuracy (%)
8	77
9	41
10	32

Many variations were studied and, generally, the 8-digit nonsignificant number seemed most accurate.

4. Not many companies continue to make the same product year after year without making changes to it. In some volatile companies, or newly formed businesses, the same products may not be made at all in a few years. If the part number system is changed, and the bill of material is placed on the computer, a rather simple study each

Figure 19 The part number should be considered the key which obtains all other data about a design. For each of use and accuracy, it should be short, consistent, and easy to use.

year will determine how many old parts are still active. When the economics justify it, old part numbers can be retired and new numbers issued for a select set of active parts.

5.   In a company with several design sections, it may be necessary to issue blocks of numbers to each of them so there will be no delay in assigning them. This implied significance should be ignored and not used; however, there is nothing more certain to make an engineer not use a part, and that is when he finds a department other than his own has issued it.

6.   No company can economically utilize classification or in some manner reduce design proliferation without a good, workable, and active set of configuration control rules. Without these, no numbering system is effective. In addition, many companies have more than one design group often in different locations. A drawing exchange system, including an adequate engineering change order procedure, is mandatory for good coordination. Use of microfilm and/or microfiche will augment this system.

7.   In those companies which do not have part numbers on some or all of their parts, a classification system should be implemented first. A decision on whether it should be used as the part number may be deferred until the quality of the finished system is checked. In any case, issuing part numbers at random in such a situation merely incorporates all the problems other companies have created over the years, particularly identical parts with different numbers.

8.   Those companies utilizing order number systems or mark numbers and group-item numbers will have the greatest difficulty in effecting change. Such systems cannot be changed overnight. Single piece-part drawings will have to be made, and a classification system will make this, rather than creating them at random, much more economical. Old assembly drawings may need to have their parts lists marked over (but not necessarily erased) and used as reference only. In some cases, the old item number may become the *find* number which equates the parts on the parts list to the physical location of the part on the assembly. In some cases, the assembly drawing must be designated as a reference drawing, meaning it does not necessarily show the exact parts needed for the assembly, but is, rather, a picture showing the general assembly.

9.   Inevitably, a problem known as *restructuring* and/or *modular design* will be faced. Other sections of this book will assist in performing this task.

PIECE PART DRAWINGS

4

Many companies are finding themselves in a position to question various drafting practices which have been established and maintained for many years. In particular, the use of, or lack of, individual piece part drawings for design documentation has become a question of serious import to implementing new techniques such as computer-aided drafting and other disciplines used in computer-aided design, computer-aided manufacturing, or CAD/CAM.

For those companies who have multipart drawings, several apparently diverse and contradictory sets of economic justifications can be presented, either to retain them or convert them to piece parts. (By multipart drawings is generally meant those which picture a complete assembly and also show the detailed parts on the same drawing.) In some cases the piece parts are detailed directly on the assembly with no separate details. Another multipart drawing has often been called a *tabulated* drawing and is a special case of the problem. While this aspect will be discussed, the principal discussion shall be with regard to the multipart drawings.

Reasons for retaining multipart drawings include the additional engineering drafting effort required, the practicality of changing shop methods, and the reusability of designs. It is the intent of this chapter to attempt to place the need for individual piece parts into proper perspective, and discuss objectively the reasons for maintaining piece part drawings or converting to them.

Some of the first complex designs documented on paper were steam engines. Many of these original drawings still exist and are creations of great beauty, drawn to scale, on linen with india ink. Part dimensions were scaled directly from the drawings and manufactured. A list of materials was placed on each drawing to designate raw materials required. Other manufacturing companies followed this example. Baldwin, Lima, Hamilton was one company which practiced this method.

Two slight, but important alterations occurred in the system. Even though it was recognized that no two steam engines were identical, many parts were common. These parts could be, and were, manufactured in batches. It was necessary, therefore, to "identify" them with a "neutral" number not tied to a specific order, machine, or assembly drawing. Thus, these parts were given *part numbers*.

Either parallel to, or because of manufacturing practices, most parts requiring "machining" in a machine shop were separated from the more complex final assembly drawing and documented as separate *machined* parts. This separation required individual piece part drawings, and unique part numbers. (Some companies have not separated machined parts from assembly drawings to this day. Consequently it is not uncommon to find an assembly drawing cut into appropriate pieces with a pair of scissors and these pieces sent with the appropriate material through the machine shop. Sulzer Brothers in Wintertur, Switzerland follows this practice.)

As more and more parts were made in batches, material control became more complex. The lists of material became longer and included a variety of different means to identify parts. As time passed government taxes on inventories also contributed to the need for minimizing quantities of parts made in batches. These became conflicting variables in the decision-making processes for manufacturing.

Purchased parts required their own separate identities (although, for a long time, "hardware" items such as bolts and nuts resisted the system of identification and their descriptions only were used). As complexity grew, and inventory control became more mandatory the list of material was "detached" from the drawing so several different departments could handle the data without receiving copies of the large "C," "D," or roll size drawings.

Final assemblies became more complex, requiring many section views to adequately picture the many parts combined. These views became difficult to interpret. In addition, those who had painstakingly learned their trade and made an "art" of manufacturing aged and retired, leaving a relatively untrained, often uninterested, group of workers behind. Work in process became larger and more expensive. and requirements for nonstandard equipment, shorter lead-time requirements, and constant "trading" of parts for one order to complete some other order created massive logistics problems. When "staging" for a finished product required all parts to be available prior to final assembly, long waiting times and pilferage or "borrowing" made the problem even greater.

As a result, products were divided into many subassemblies, each requiring their own drawing and also individual part numbers. Sometimes identity of these subassemblies were made using job numbers tied to the final customer order number. This created an inability to recognize when identical assemblies were in process at the same time. Keeping track of these drawings assemblies, and, often, detached parts lists required a complete set of new rules. The rules thus established and in effect today are now known as *configuration control*.

While the natural expansion of these methods continued, production and inventory control passed through stages of development to control inventory known as *mass release*, and *order point control*. Each of these systems became effective only as each part and assembly were identified individually with part numbers.

Strangely, a completed weldment also considered a nondismantable assembly continued to be considered a *piece part*. Part of this philosophy was based on the fact that none of the parts in the weldment were considered to be service items, and part was based on a preconditioned belief that welded parts were "cheaper" than machined, and rarely reusable in other designs.

Ironically, because machined parts had long been considered separate entities, crude but usable retrieval methods had been designed. Since retrieval methods for such parts did prove the ability to reuse them, it was common knowledge that "common" parts existed and were practical in this category. The expense to manufacture machined parts was more thoroughly documented, also giving proof that producing machined parts in batches was more economically practical.

Weldments and their associated parts then became the last segment of manufacturing not to be identified with unique part numbers and separate drawings. Only companies with obviously high volumes of weldment production such as Caterpillar Tractor Co., General Motors, Control Data, etc., documented fabricated parts as individual items.

In most companies that do not now have piece part drawings, much emotion has been evoked in sidestepping an objective analysis of conversion. Often heard are such questions as, "why should we have to spend engineering hours documenting on a separate drawing a simple 4 in. × 6 in. × 1/2 in. rectangle?," or, "we never make the same assembly!." It is time to look objectively at the need for such part documentation.

Several important reasons exist for requiring piece part drawings. The following items will be discussed individually in some detail:

1. Clarity of fabrication details
2. Economical part design
3. Documentation changes
4. Part reusability
5. Design retrieval
6. Elimination of duplication
7. Raw material requirements
8. Principles of manufacturing
    a. Routing requirements
    b. Tooling requirements
    c. Group technology
9. Family drawing preparation
10. Computer-aided drafting
11. Purchased parts

In discussing these points, additional drafting costs must be considered, the percentage of "simple" parts in a fabrication, as well as the means for documenting them. Furthermore, design retrieval methods, costs for creating new designs, and other means for documenting parts should be considered.

Let us look first at a rather simple example of the kind of drawing we will discuss.

Figure 20 shows a fabricated assembly with most of the detailed dimensions eliminated for clarity. It is implied that all dimensions for making the part and the assembly are on the drawing. Since each part is not separated one from another, two methods can be used to "identify" the parts. Often called *balloon, bubble,* or *find* numbers, they are used to locate each part and are normally attached to the assembly drawing to create a part identity. Thus 606442-001, 606442-002, and 606442-003 become the piece part identifiers or part numbers, while 606442-000 becomes the part number for the assembly. Some companies also attach meaning to the balloon numbers by making the line number on the detached parts list equal to the balloon numbers. This, then, is a typical drawing of the type we will discuss.

Let us next look at the results of preparing a new design quite similar to this one with only a minor alteration.

For example, as shown in Figure 21, let us change the overall length from 18-1/4 in to 21-1/4 in. Normal procedure is to make a *sepia* or reproducible copy of the original drawing, change the appropriate dimension, and assign a new assembly part number. Immediately three completely new parts and a new assembly have been created. Two parts which are identical between the two assemblies have been "hidden" by being assigned new part numbers using the new assembly drawing number as, 606999-001, 606999-002, 606999-003, and 606999-000.

Even if these two assemblies were placed onto the same bill of material, there is no possible way of finding identical parts without careful examination of each one. Thus each part will be routed, planned, and issued to manufacturing as different parts.

With this brief introduction to the subject of piece part drawings, we will begin to examine each point previously defined.

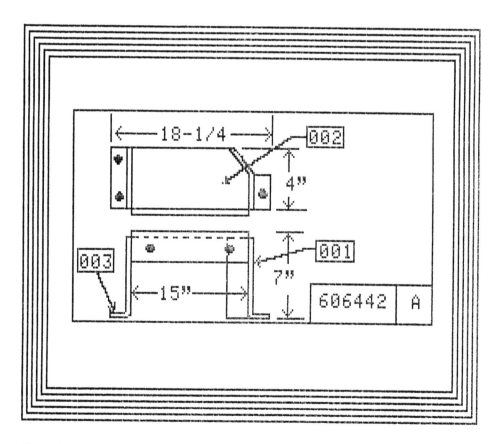

Figure 20 This multipart drawing can be used as an example of the problems incurred in not drawing piece parts.

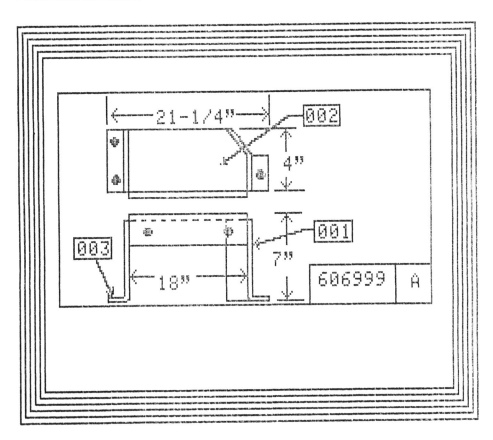

Figure 21 Slight changes to a multipart drawing create major documentation changes and eliminates the use of identical parts.

CLARITY OF FABRICATION DETAILS

With the advent of a computer-generated bill of material, it became mandatory to identify each part with a part number. However, the method used was to give official recognition of the need to combine the assembly part number with find numbers to create the part number, as shown in Figure 22. The philosophy and system was not changed.

The result of this incomplete change inevitably created problems. For example, when the steam engines were first manufactured, master craftsmen interpreted drawings. Conscientious, hard working, with great pride in their work, they "hammered out" the iron. Where, from experience, they discovered identical or similar parts, then jigs, templates, or fixtures were made to recreate parts faster for second and third useage. "Master sargeants" of manufacturing, they rarely told engineering when identical parts passed by them. After all, it was their job to make the parts required, not question the "officers" (engineers) on how they were documented. Fit up and tolerances were not critical in the early days of manufacturing. Welding techniques and preparation were almost unknown.

Those days have disappeared. Master craftsmen have become a host of specialists.

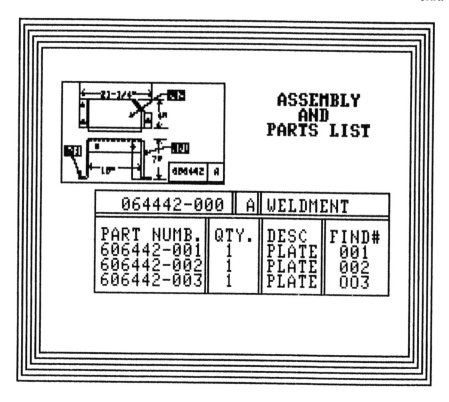

Figure 22 Mechanization of engineering documentation is enhanced by removing the list of materials from the assembly drawing and making a detached parts list. The parts list also becomes the collection point for the piece part drawings which were formerly collected on a multipart drawing.

Many alternate manufacturing techniques have appeared. Many different materials have been created. Tolerances have become closer. Welding has superimposed new manufacturing requirements. Designs have become more complex. People have become less skilled.

Figures 23 and 24 are simple fabricated assemblies. However, questions arise. Should the holes be prepared on the piece parts prior to assembly or not? What are the tolerances of the piece parts before they are assembled? (The parts shown in Fig. 23 are extracted from actual drawings.) Confusion arises as to the meaning of the engineering notes, the almost impossible, or, at least, expensive requirement to drill holes through two unstable parts welded to a third part, seems impractical. The part shown in Figure 24 is a much simplified example of the complex dimensioning required to detail all parts of an assembly without drawing each individual part.

Analysis of this type of assembly occurs three times. When the engineer first designs it, he has a probable method of assembly in mind. However, this sequence is not normally found on the drawing. The industrial engineer analyzes the drawing the second time for tooling, templates, and routing sequence. If he or she assumes the holes will be drilled or punched after assembly, the part templates will not provide for them. A third analysis takes place when the layout people redetermine the templates needed, and make *flat layouts* of the parts. In all cases the assembly must be "taken apart" mentally. Each piece part is pictured separately.

The bill of material shown in Figure 25 is a confused mixture of parts for an assembly, with part numbers, with bubble numbers and no numbers. To stage a complete assembly requires collecting a variety of parts, some with identity, some with none. In addtion, at least one fabricated part has been defined on a separate drawing, and thus it must be collected separately from the place of manufacture, the dimensions must be checked to see if they match those on the assembly drawing, and finally, it must be put together. Writing notes explaining the assembly procedure on the drawing would only serve to eliminate possibly more economical methods known in the industrial engineering field.

Figure 26 shows the same parts which were used to make the assembly shown in Figure 20. Although the dimensions have been omitted for clarity, it seems obvious they will not confuse the fabricator. Rather, the parts would be clearly marked, and the requirement for drilling or punching holes in the parts prior to assembly clearly and explicitly shown. It should be noted that, since the holes are shown in the parts

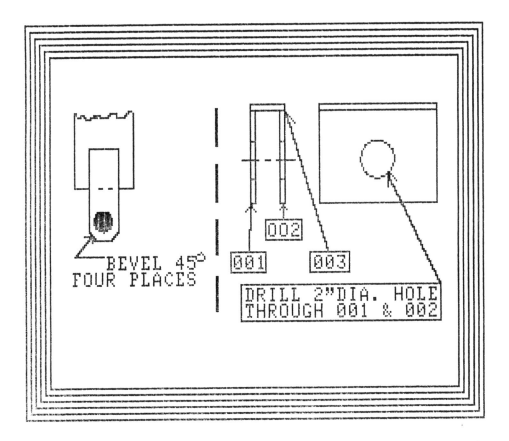

Figure 23 Clarity of manufacturing is enhanced if separate piece part drawings are made, and the operations required at the part and assembly level segregated.

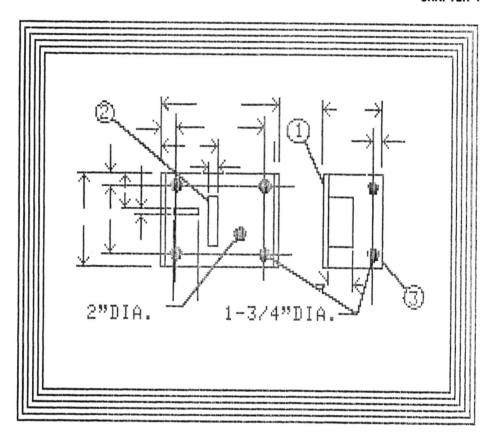

Figure 24 A complex assembly can be made even more difficult by requiring both piece part and assembly dimensions on one drawing.

prior to assembly, the industrial engineer can make a judgement about this method, and change the drawing before fabrication actually begins. Clearly, the use of piece parts also makes it possible and wise to have close coordination between design and industrial engineers to set logical fabrication practices for parts such as these, so that subsequent designs will be issued following mutually agreeable practices. No longer would each planner and industrial engineer create their own separate ideas as to the best fabrication practices.

The conclusions which can be made from these simple examples include:

1. Manufacturing operations on the assembly versus operations on each piece part can be clarified and defined economically.

2. Individual dimensions of parts can be made clear. A combination of assembly dimensions and part dimensions will not cause confusion.

3. The assembly drawing will contain notes for the assembly operation only, not a series of confusing notes for both the parts and the assembly.

4. Since each part has an individual drawing, all other parts with identical material from other assemblies may be grouped together and routed to the cut-off or shear-

ing and burning section of the fabrication shop thus economically utilizing the raw materials available in a logical sequence.

5. Although it is claimed that drawing piece parts is more time consuming, it is a fact that every dimension required to make the piece parts on a multipart assembly must appear on such an assembly somewhere. Consequently, there is, in reality, no additional dimension required. Although the lines of the parts must be repeated, planning for multiple layers of dimensions to describe both the assembly and the part dimensions has been eliminated, and implementing "family" or "format" drawings for common shapes of parts eliminates the need to make even the drawings for many parts.

6. For those "simple" parts which are rectangular or square and require no operations other than shearing, a simple typewritten drawing is sufficient.

7. Since each piece part may have different operations on them, such as drilling or punching, notching, beveling, bending, etc., individual part drawings allow each part to flow through the machine or fabrication shop individually with an identity on the part as it travels.

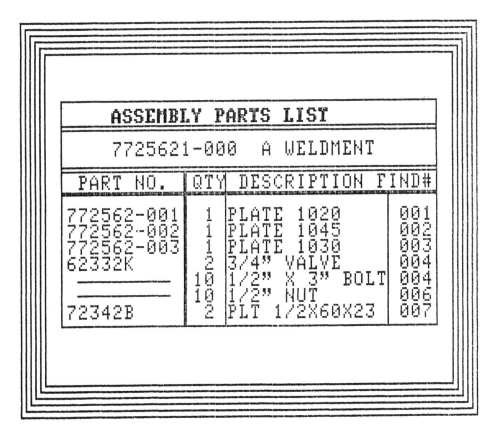

ASSEMBLY PARTS LIST			
7725621-000 A WELDMENT			
PART NO.	QTY	DESCRIPTION	FIND#
772562-001	1	PLATE 1020	001
772562-002	1	PLATE 1045	002
772562-003	1	PLATE 1030	003
62332K	2	3/4" VALVE	004
————	10	1/2" X 3" BOLT	004
————	10	1/2" NUT	006
72342B	2	PLT 1/2X60X23	007

Figure 25 If mechanization of the bill of material and parts list is not accompanied with a good part number system, the results will be confusion in the identity of the parts. No bill of material can be mechanized without a part number system.

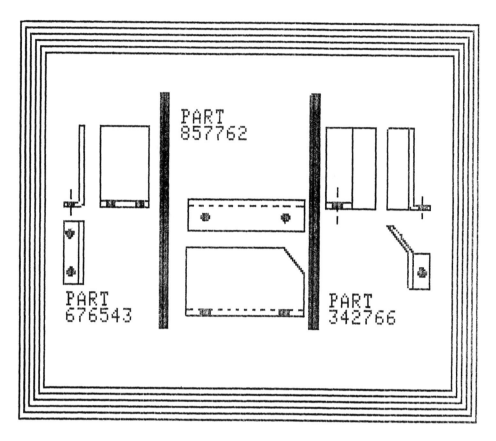

Figure 26 The piece part drawings shown here help to clarify both their manufacture and the assembly from which they were taken.

These are important and logical reasons to restudy the need for individual piece part drawings. We shall turn next to another reason for piece part drawings, economical design.

ECONOMICAL PIECE PART DESIGN

Figure 27 shows a bent bracket with a center reinforcing bar. In the illustration, the reinforcing bar shape has been grossly exagerated to show the problem. In the actual drawing from which this illustration was extracted, the drafter made the two lines for the reinforcing bar appear parallel.

The economics of manufacturing this part are obvious when the part is extracted from the assembly. The "wrong" method shows a part which requires four operations to shear it to the proper dimensions. The "right" method shows a part which may be sheared from a bar requiring only two operations, and also allows for using a standard raw material. Hundreds, perhaps thousands of parts in a manufacturing company are designed in this manner, thus creating this type of problem. Since such parts have been designed for many years, the manufacturing personnel have come to accept this expensive design. It is also disarming in that such designs appear over a long period of time and are not released in clusters. It is also true that the cost differential is not excessive because it is a simple part. When multiplied over thousands of poorly designed parts, however, the saving potential is enormous. It is probably important to remember:

There Are No Million
Dollar Savings
There Are Millions
of
One-Dollar Savings

Returning to the example shown in Figures 20 and 26, the piece parts shown are not only relatively simple parts, but are also very likely to be used again in other similar assemblies. Each part might be made from different raw materials, and different sizes of material. It seems obvious that issuing these parts as a group will cause a series of parallel events to occur which are time consuming and expensive. Since other assemblies are being manufactured at the same time, but no means for collecting dif-

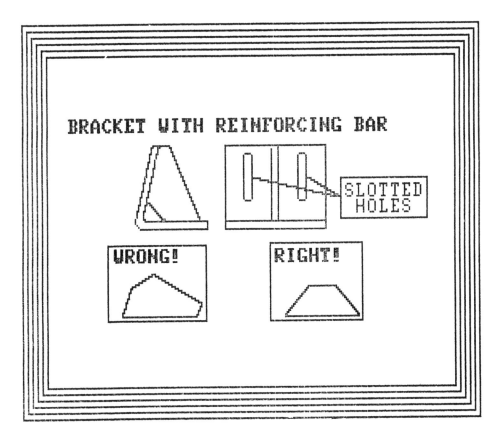

Figure 27 A more economical design can be prepared if the consequences of the design are pictured. Design of piece parts make this possible.

ferent parts with identical material is available, each separate assembly must be staged individually by requesting suitable material for each part from steel inventory and cutting the necessary quantity for each part. Normally the remaining material is returned to stock only, perhaps, to be reissued for the next assembly. Obviously, common sense will allow foremen to "manually" attempt to group such jobs in the shop, but the resulting manual correlation and routing of material and parts can be less than optimum, and subject to material loss, excessive raw material racks in the manufacturing facility, and many extra manual systems to keep track of the activities. Every action that a good foreman may take to optimize production can be greatly alleviated with a good piece part numbering system, and corresponding piece part drawings.

Many parts that have re-entering cuts, and curved portions require burning templates. These templates are expensive, and also require an identification system and storage facility of their own. Even with numerically controlled burning equipment, the paper tapes or other methods used to prepare the parts require an identification system. One company estimates over 25% of all burning templates are duplications. Thus, cost of their preparation combined with difficulty of retrieval adds unnecessary costs to manufacturing. Often a "solution" for routing parts on a multipart drawing is conceived by preparing a routing for the assembly only and then listing by operation each individual part that requires such an operation. Figure 28 is an example of this "solution." Let us look at some of the problems this system creates.

The manufacturing process sheet calls for a fabricated assembly part number 6675432. The shop order number for a quantity of 20 parts is 843313. The process sheet is composed of the following data:

Operation sequence number

Operation number

Department number

Machine number

Machine set-up time

Part run time

Three operations appear on the sheet, namely 63 angle shearing, 160 mark burn and clean, and 170 punching.

Since most, but not all of the parts in the fabrication have been assigned part numbers, they appear, when applicable, on the left side of the listed parts. Where no part number has been assigned, a blank space appears. Since all parts do not have the same operations on them, all part numbers are not listed under each operation. Not only are all numbers not listed for each operation, but some disappear and reappear again depending upon their configuration. For example, part number 434425 appears in all three operations, part number 434222 appears in the first and third operation, part number 438118 appears in the first and second operation, and part number 438119 appears only in the first operation.

The first operation is supposedly an angle shear operation, but plate with a different chemistry is also listed. This actually requires a plate, not an angle shear, and requires a separate operation.

The size of angles vary, so that, even though they might be sheared on the same

```
O-O-O-O-O-O-O-O-O-O-O-O-O-O-O-O-O
O    MANUFACTURING PROCESS SHEET    O
O       PART NUMBER--667543-------    O
O       ORDER NUMBER--- 843313----    O
O       STANDARD QUANTITY---20----    O
O-O-O-O-O-O-O-O-O-O-O-O-O-O-O-O-O
```

```
SEQ  OP.DEPT MACH  SETUP  RUN
NO.  NO  NO   NO.   TIME   TIME

05   63   93   3924    .15 .650
                 ANGLE SHEAR
```

                    FOLLOWING CUT FROM A100

```
   434425   (01)  1.50X1.50X.188X94.00
   434222   (01)  1.50X1.50X.188X59.438
   433441   (02)  1.50X1.50X.188X23.50
   438118   (02)  1.50X1.50X.188X23.50
   438198   (01)  1.50X1.50X.188X13.438
   437748   (02)  1.50X3.00X.188X 9.25
   438114   (01)  1.50X4.00X.188X 6.25
   438119   (01)  1.50X4.00X.188X 6.25
```

                    FOLLOWING CUT FROM A66

```
   466223        .25X24.88X110.38
                 .25X24.88X110.38
                 .25X24.88X110.188
                 .25X21.75X 21.75
```

```
05   160  802   4131     .10 .500
           MARK, HANDBURN AND CLEAN
```

```
   434425                      FOR 65
   433441                      FOR 65
   438118                      FOR 65
   437748                      FOR 802-65
   438114                      FOR 65
   438119                      FOR 802-65
```

```
05   170   93   3624     .25 .500
               PUNCH
```

```
   434425
   434222
   433441
   438114
```

Figure 28 The process sheet shown here is used to manufacture parts for a multipart drawing. Notice how part numbers (in the left-hand column) are repeated over and over again under each operation. The result is a second bill of material which must also be changed when the design bill of material is changed.

machine, it requires moving three separate raw materials to the shear from the yard plus possibly two different plate sizes.

Under the second operation, the "for 65" and "for 802-65" implies a note or instruction telling the worker the next operation needed on the parts, an unnecessary and possibly contradictory instruction.

Last, but not least, every instruction and operation could have been followed just as easily if each had listed separately on individual process sheets by part number. Obviously, the operation sequences and other data about the operations would have had to be duplicated, but a mechanized process sheet system from which this example was extracted could be designed to eliminate manual duplication.

Had the parts been separated, additional parts with the identical material and/or operations could then have been collected together by similarities not by a collection of parts in an assembly as shown here.

Figure 29 is the picture of the parts which are used in the assembly shown in Figure 23. In this case, however, examples of the piece part configurations have also been shown. Notice also that additional holes have been added to the assembly. The question arises, if the piece part drawings are not prepared, will the bolt circle-type holes be drilled after final assembly or in the piece parts? Should all holes be machined in

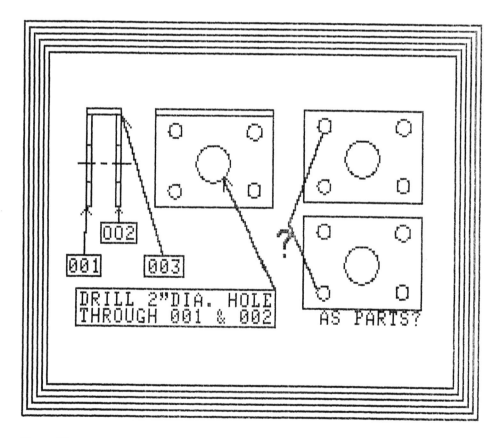

Figure 29 Piece part drawings allow the planner in manufacturing to see how the engineer planned the design including the critical dimensions and tolerances.

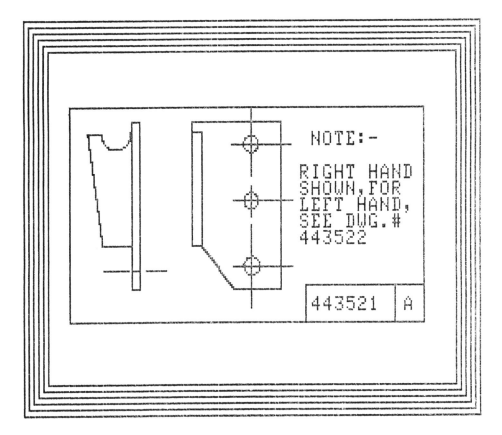

Figure 30 Piece part drawings also allow economical design of the assemblies for which they are manufactured. Such problems as symmetry may be alleviated.

the parts prior to assembly, or after final assembly? These, and many more questions are posed in complex assemblies of this type.

Figure 30 shows another common problem with multipart assemblies. Often it is necessary to build a "right" hand as well as a "left" hand weldment. In this example, showing the left-hand assembly, it is apparent that the actual part or parts are identical until they are either bent or welded together. Thus, lack of symmetry should not increase the cost of manufacturing this assembly as it inevitably will should the identical configuration for the parts not be recognized.

We must thus come to some inevitable conclusions:

1. Economical part design is based on clarity of dimensions, and an allowance for alternate manufacturing methods.

2. If an assembly may be compared to a molecule and a piece part to an atom, the restrictions of manufacturing are far less expensive by preparing the "atoms" first and then assembling them into a "molecule." After all, 98 atoms are used to make millions of molecular combinations.

3. Single piece part drawings allow far more flexible routings, including the ability to "group" parts made from raw materials, or parts with similar or even identical shapes.

4. Clarity of part dimensioning avoids incorrect conclusions as to their fit up, and even how they should appear as a finished part.

DOCUMENTATION CHANGES

If we rescan Figures 20 and 21, it is apparent that raw material specifications should appear on the drawing. Even though we might add them to make a more complete drawing, it is very often required on the parts list as a portion of each piece part specification. In either case, let us assume the raw material specification is changed, and the old material is never again to be used. Figure 22 is the typical parts list for the assembly. Should the raw material be added to the parts list, a new part number must be added which replaces the old part number. However, since the old part number is composed of the find number and drawing number, the find number must also be changed, since the parts are not interchangeable. The same condition occurs if the raw material appears on the drawing only, since a new part number is required in both cases. Confusion exists. Today, unfortunately, this type of change is not usually made. Therefore, two different parts with the same part numbers will appear in the product. Only a note on the drawing will document that the change has been made. Any mechanized system utilizing part numbers will not be able to support documentation changes such as these.

It should be noted that if the raw material appears only on the parts list, this means two documents are always needed to make a part, the drawing and the bill of material, or parts list. Obviously drawing content plays an important part in the documentation of design.

Figure 25 is an example of another documentation problem since it a mixture of parts which have been changed to piece part drawings and others which are documented as multidrawing parts. Some fabricators might conclude that those parts not identified with a part number have no operations to be performed on them, which is, of course, not necessarily true. In addition, a combined engineering change to a piece part with no part number and a change to a part with a part number are each handled differently, again causing confusion in the procedure for making such changes.

This mixture of part documentation also holds surprises for the fabricators, since they may receive the documentation necessary to make a multipart assembly only to find—to their surprise—that they must wait and order additional drawings of the single piece parts which have been documented on separate drawings with their own part numbers.

Many other examples of problems encountered when making engineering changes could be recorded. Massive proof of a problem's existence and consequences would probably lessen the effect of the few simple examples shown here. However, product liability, spare part provisioning, manufacturing changes, product reliability, and the ability to reproduce the same product again rely on sound documentation. Multipart drawings do not appear to enhance these requirements.

PART REUSEABILITY

Perhaps one of the best ways to illustrate what part reuseability means to productivity can be grasped by a graphic example. Picture a drafter sitting at a drafting board with a "C" or "D" size drawing on the board which has on it a complete assembly of some fairly complex weldment. After studying the drawing, the drafter notes that, with the exception of two dimensions, it is the precise design needed. Perhaps there are 30 different parts on the drawing and 28 of them require no changes. The drafter has three choices. Recognizing the number of parts which are identical, it would be possible to make piece part drawings of each identical part, identify them, and revise the old design to show the new parts, both for the new design and for the old. It is a foregone conclusion that the drafter's supervisor would never hear of such a process because of the time involved. Consequently this choice is discarded.

The drafter can take out a new piece of paper and redraw the new assembly in such a manner as to show alterations to the old assembly to make the new. This choice requires too much time also, and, in addition, creates a new problem in that the new assembly drawing must refer to the old and proclaim the new, in a "same as but" documentation. (It has been found that this type of documentation almost inevitably creates another sequence of events whereby the next drafter finds the altered drawing and makes a "same as but" drawing which refers to another "same as but" drawing, etc.) The interpretation required in the shop to make such an assembly requires the patience, logic, and time of a superperson. At any rate, this solution is discarded also.

The only solution left is to make a copy of the original design as a sepia or with some other reproducible method, erase and change the offending dimensions, attach a new part number to the new assembly drawing, and — immediately issue 28 identical parts under 28 new drawing-find numbers.

Since this type of activity snowballs over many years, one example at a time, the apparent conclusion is "we never make the same part twice." It is a mistaken conclusion with little or no means to disprove it. On the other hand, once such an example or examples are found, another argument for continuing the practice is that it is cheaper to make the same part over under a new number than take the time to document them as piece parts. Let us take an example of a simple rectangle to illustrate the inaccuracy of such a conclusion.

Figure 31 portrays the cost for making the "simple" rectangle. The actual drafting and designing cost is $5, or about 15 minutes. Upon release of the part, a routing slip and other documents are prepared in manufacturing. It is also entered into a mechanized system. The conservative estimate of $3, or about 10 minutes seems more than reasonable. The actual manufacture of the part, including material costing a total of 50 cents also seems conservative. The total cost for the manufacture of such a part including a 50% burden is thus $12.75. If this part is made again under another multipart drawing, once again the cost will be $12.75, and will continue to be as long as the part is documented under different drawing-mark numbers. As illustrated, the total cost for making such a part twenty times is $255. However, if the part had been documented as a piece part when first designed, making the part again and again would have eliminated all documentation costs listed and total cost for twenty parts would have been $31.75, including the cost for documenting the first part. This is a substan-

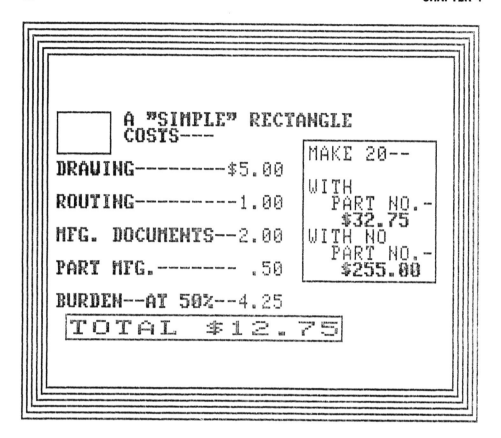

Figure 31   Piece part drawings make it possible to reuse an old design. Simple designs aren't inexpensive. Once even a simple design is documented, money can be saved when it is reused.

tial savings when extrapolated over the hundreds or thousands of parts manufactured yearly in a multipart drawing company.

Lest anyone believe the duplication problem does not exist, Figure 32 is a list of duplicate parts taken from a sample of 1000 multipart drawings in a company whose major products were generally repetitive. The largest set of duplicate parts documented is 41, but we can be sure a rigorous study would find more. Remember also, most duplicate parts are not as simple as this example.

We must never forget that, no matter how inexpensively a part can be documented, the customer must pay for the entire cost if it is used only once. If the original design can be used just one more time the cost of such documentation can be reduced by 50% in terms of customer cost. Conversely, if a system is designed so that it is difficult or impossible to retrieve duplicate parts, the duplicate parts will never be found.

Figure 15 showed a method which has often been used to assure part reuseability. The drawing described previously is called a *tabulated drawing*. Many companies have hundreds of them with the first drawing recorded as long as 40 years ago.

In studying Figure 15, we see that 31 separate drawings of a straight bar with five equally spaced holes have been documented over 27 years. In all, 12 different dimensions have been tabulated. To any designer with common sense (and a good memory), this document represents a means for finding such parts. Unfortunately, this theory simply is untrue. As we saw in Figure 16, parts which have been extracted from the tabulated drawing show what parts have really been documented. We have already discussed these variations.

Lack of discipline and interest inevitably make this type of "short cut" unuseable.

In some cases, a tabulated drawing is a direct reflection of some vendor catalog, encompassing every possible variation, even when only one size or capacity has a known use. One enterprising engineer was able to make a matrix of 2400 parts on an A size drawing. Such a waste of part numbers and proliferation of designs create a system showing little discipline and a lack of respect by manufacturing personnel for drawing documentation. The disadvantages of a tabulated drawing may now be apparent. There are several reasons for discarding them as a means for recording reuseable parts. Let us summarize them again for emphasis:

Duplicate entries

Wastes part numbers

Confusing to the fabricator

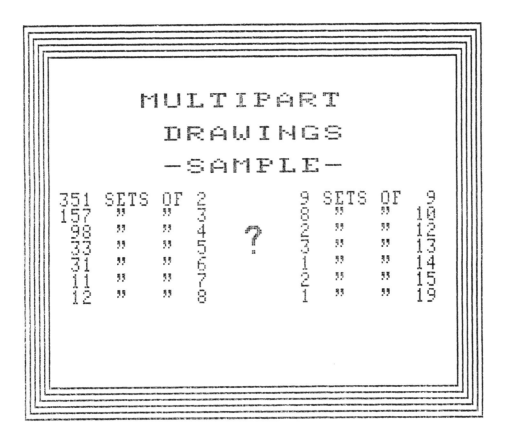

Figure 32 Most companies sincerely believe their designers don't make duplicates. Here is an example which shows quite the opposite.

Revision changes difficult

Error in dimensions look-up

Requires shop special handling

Unlimited (not preferred) items

Implies inventory requirements

Often requires other documents

Creates unnecessary paperwork

Does not assist in design retrieval

The picture does not equal the part

How many piece part drawings are required on the average to meet drawing requirements? A sample of 70 assemblies was examined to determine this. An average of 6 parts per assembly were required, or 420 parts. Of this total, 188 or 41% had been previously identified as standard parts; 64 or 15% could be documented as plain parts using a typewritten drawing; and 186 or 44% required new drawings. On the average, then, one new typewritten drawing and 3 part drawings were needed per assembly. An examination of the piece part drawings required showed none with such complexity as to radically alter drawing time. In fact, removing piece part dimensions for the assembly drawing allowed simpler and faster assembly drawings, since it was not necessary to preplan lines of dimensions to clearly show part dimensions and assembly dimensions on the same drawing. In a later chapter we will discuss design retrieval in detail. However, let us explore some of the reasons for a retrieval system as it relates to the economy of piece parts.

If we have in any way begun to prove the need for piece part drawings, it is then also necessary to make certain each one of them may be used as often as possible. Consequently, we must examine the best method to achieve such a goal, utilizing design retrieval systems.

DESIGN RETRIEVAL

Once piece part drawings are the policy of company, such drawings may be prepared yearly. It is ironic that the labor for preparing these drawings becomes wasted if they are not easily available when needed for reuse. Consequently, a method for retrieving these drawings rapidly and economically becomes essential. If the mention of any system causes an emotional response by engineers, design retrieval evokes the loudest outcry. The picture conjured up by an engineer is one where creative design is repressed, and mountains of paper work are created just to find a part far more easily drawn than retrieved.

Many of the larger companies such as Caterpiller Tractor Co., Boeing Aircraft Co., Control Data Corporation, American Standard, and more have had design retrieval systems for years and they have made substantial savings in the application of retrieval. For example, Caterpillar Tractor Co. claims to have no duplicate parts in existence. In addition, 30–40% of all parts in their new D-10 tractor were originally designed for their D-9 tractor. This implies high competence, and good discipline in the use of their classification and retrieval system.

Let us first define what classification and design retrieval should mean as a system:

Classification and coding   A system in which individual items are grouped together by virtue of their similarities, and then separated by their essential differences into "families" characterized by their features or function.

Two key words must be associated with a design retrieval system. It must be possible to, first, "identify" a design in such a way that it can then be "retrieved." Five general methods have been devised to design a retrieval system. Let us look briefly at each method.

BLACK BOOKS

Every engineer, designer, and draftsman worth their salt has a black book. Indeed they may have black books from other companies for which they have worked. Unfortunately only that individual is privileged to look into it and find parts and assemblies to use. Thus, 100 designers and draftsmen with 100 black books make a company look like a series of fiefdoms which cannot and, perhaps, do not want to communicate with one another. The antithesis of good economical and safe design is a set of black books.

RANDOM CODING

Random coding utilizes the skill of the classifier who identifies various shapes of parts as they are received, and rather arbitrarily applies a code number to each new general shape. An index is used to categorize and identify the various designs, with a picture of the variations usually mandatory to augment the verbal description. Since no orderly procedure is preplanned prior to the classification effort, retrieval becomes difficult for the specialist and almost impossible for the novice due to the vast quantity of categories in the index. Even if the categories are preassigned and separated by some numbering system, an inordinate number of designs will inevitably fall into the same section, either making it difficult to scan the total population of parts, or forcing the coding part of the system to fail, thus making a new category for essentially the same parts.

NAME CODING

Probably the oldest form of classification is use of the part name with assorted adjective modifiers. Often it is difficult to determine whether the description should consider the function, shape, size, color, or some other characteristic of the part. If function is used, it is often found that identical parts have different functions. If name only is used, major disagreements arise as to which name should be used. Minor differences in designs may require different code names for retrieval, but finding new nouns and adjectives to describe such differences is difficult if not impossible. Figure 33 is a list of names which can be used for a simple round part. Obviously if all of them should be used retrieval would become difficult.

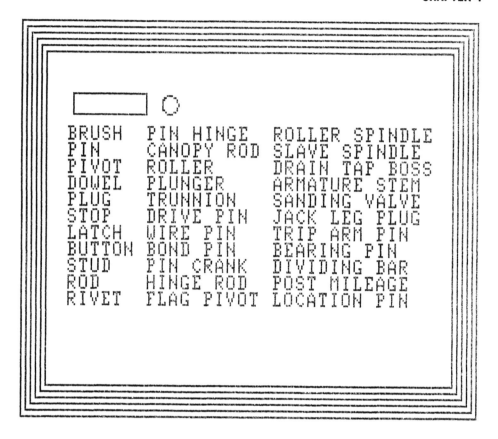

Figure 33 Retrieval using the name of a part is almost impossible except under specific conditions. Thus, name coding is a dangerous means to retrieve designs.

POLYCODES

Growing in favor and increasingly implemented, retrieval systems using *polycodes* are becoming more popular. Certainly in the manufacturing area where machining characteristics need to be identified they are serving their purpose. The definition for a polycode system follows:

> *Polycode system*   A set of independent sequential codes, each digit or set of digits describing individual characteristic variations. Polycodes are often described as *feature* codes.

Great care must be used to implement a polycode system or the code number will be extremely long, and also full of characteristics which do not apply to a large group of the parts thus classified. Retrieval of designed parts is not the strongest part of such a system.

MONOCODE

Many of the most efficient design retrieval coding systems are *monocodes*. The logic used in such a retrieval system has been finely honed over the years to follow the normal instincts of an engineer who mentally describes a part. The definition of a monocode follows:

> *Monocode*   An integrated progressive coding system in which each digit holds information which qualifies the information contained in the previous digit. Each digit may contain more than one *descriptor* or *characteristic*.

Regardless of the system used, the following criteria should be met:

It must be all inclusive

It must be mutually exclusive

It must have logical sequential decision trees

Permanent descriptors should be used

Frequency of part types is essential

A predesigned discipline is important

To be successful design retrieval needs the support of good discipline in the engineering department, including restrictions on the issuance of new part numbers before a search of the system for an existing part. Retrieval should take no more than a minute if an engineer is to feel comfortable with the system.

The rewards of a good classification and design retrieval system in combination with piece part drawings, however, are great. A few of them are listed below:

New designs by modification

Development of modular design

Make–buy decisions

Parts standardization

Use of "family drawings"

Defines preferred, nonpreferred parts

Establishes training program for engineers

More uniform costing, routing, purchasing

Shorter new product lead time

Elimination of duplication

Often a manufacturing company installs classification and design retrieval systems, breathes a sigh of relief, pays the bill and then—conducts business as usual. Later, when asked whether the system is performing as planned, the response may be vague answers. Perhaps the code books and sets of rules are over in a corner gathering dust. Certainly many classification systems have been misused and not used. Elimination of duplication is one of many byproducts of the system which pays magnificent dividends to the serious, conscientious user.

Figure 34 is an example of a simple part sometimes called a "spigot." The group of drawings that were documented in this family of parts were 27 in number. Of course, this supply of parts should be thought of as a repository of designs to which the engineer and designer can go and interrogate. In addition, however, there may simply be too many parts in the group and many could be eliminated. Sure enough, a study of the parts in this group made by one company found the following reductions could be made:

11 Outside diameters to 5   4 Hole sizes to 2
12 Inner diameters to 5   5 Counterbore sizes to 2
14 Lengths changed to 3   9 Descriptions to 1
13 Internal diameters to 5   8 Sets of tolerances to 1
2 Hole patterns to 1

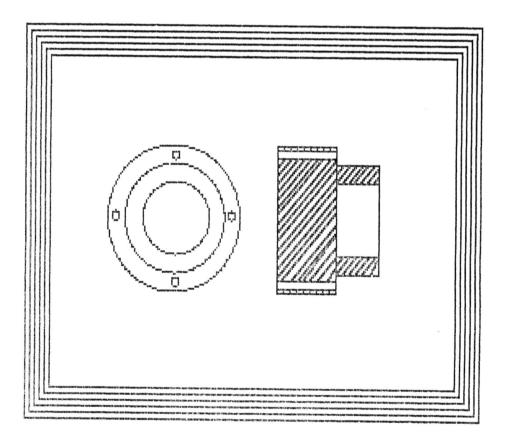

Figure 34 Once parts have been classified into families using similar permanent features, they may be studied to eliminate design redundancies.

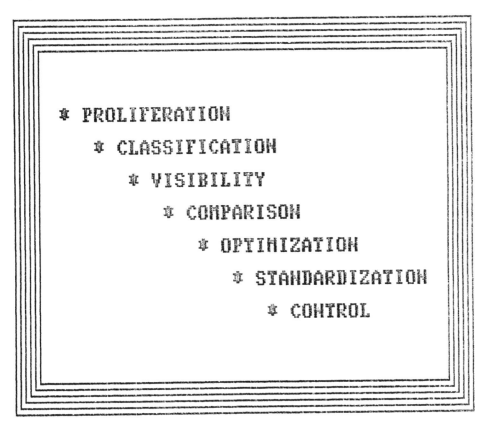

Figure 35 A classification and retrieval system is a prime tool for creating an orderly, systematic, and logical design department.

A logical survey of families of parts will find many such possibilities. Even should there be no standardization possible, there will surely be those parts which are preferable over others. This will serve to steer the engineer to select parts with the best and/or the most economical characteristics. Proliferation of similar parts is no longer an economical choice. Through this type of effort, consolidation of design is possible.

Figure 35 summarizes the objectives of a good classification design retrieval system:

*Proliferation of Parts Should be Classified for Visibility in Order to Compare
and Thus Optimize to Allow Standardization and Thus Regain Control*

It is just no longer economical to continue making new parts and assemblies whenever it is thought necessary. The economics of proliferation of parts does not justify it. Perhaps another way of saying the same thing is to think of taking chaos and making it orderly so that it is possible to have orderly chaos.

RAW MATERIAL REQUIREMENTS

Many companies have a need to summarize their raw material requirements for the parts they manufacture. One of the prime reasons for mechanizing the bill of material system was to be able to perform such an operation.

Suppose a company has 100 products with 10 to 15 levels in the bills of materials, handling 5000 assemblies and perhaps 10,000 piece parts. Summarizing the total requirements of the parts and their corresponding raw materials is difficult if not impossible on a timely basis. If material requirements planning (MRP) is used to perform this function, the task may be performed smoothly, accurately, and rapidly. However, material requirements planning assumes part numbers on individual piece parts, and part numbers for each raw material. Then all parts required may be summarized and converted into the tons of each raw material required. When considering that material requirements planning is used to forecast total requirements of these parts and raw materials for 52 or more weeks in advance, the size of the problem can be seen to be substantial. Only a good mechanized system can perform this task readily. We must therefore consider that accurate and timely raw material requirements depend upon accurate documentation of the piece parts for which the raw material is required.

Not only raw material, but also castings and forgings require adequate documentation. Rough forgings and castings should have their own individual part numbers so they can be used to make additional finished parts. Thus, a higher volume of rough castings and forgings can be manufactured. In one company, a sample of 400 sheaves or pulleys showed 45% of them could be altered slightly so one casting could be used for more than one finished part. For example, one rough casting could be used to make 32 finished pulleys.

Thus, some of the rewards of a good piece part documentation system occur in maximizing inventory, and a rapid calculation of requirements.

PRINCIPLES OF MANUFACTURING

It is of little value to design a part or an assembly under any system if the part is not manufactured. Too often we forget that engineering drawings and their associated documents are communications to others. If this communication is difficult, incomplete, or ambiguous, manufacturing costs increase, or the lead time of the product is penalized. Let us review some of the documentation we have presented, and summarize some of the possible results.

Figures 20 or 21 are typical multipart drawings. Figure 25 is a typical parts list for such a multipart drawing. Figure 28 is a typical process sheet or routing for a multipart assembly. The routing sheet in this example is also the work order needed to make the assembly. If, for a moment, we consider these documents to be all for the same assembly, we can list some startling conclusions:

1. Every operation required to make each piece part appears on the process sheets and are equal to the same effort had each piece part had an individual routing prepared.

2. Rather than describing detailed operations for individual piece parts, identical

operations are described, and those parts meeting the requirements of the operations are listed by part number under the appropriate ones. As a result, part numbers appear and are repeated for each operation the part requires.

3. As noted before, each piece part may not require all operations listed. Thus, each part and its individual operations must be traced meticulously through the operation sheet. In addition, diverse materials are shown under one operation.

4. In addition to the many times the part number must be repeated to define its individual operations, two additional lists of the same part numbers are usually maintained, one showing the list of parts to be delivered to the next department, and the other a list of drawings required to make the parts. Of course the parts list shown in Figure 22 must also be maintained.

    Do you think it is possible to have all of these documents in complete synchronization at the same time?

5. It would appear to be possible to make some economies by cutting all parts which are made from the same raw material at the same time. Individual process sheets for each piece part on this assembly and all others would make this possible. Unfortunately, the single process sheet shown, which covers all parts for just one assembly has not been designed to group all parts for all assemblies having the same raw material.

6. If an engineering change occurs, in order to change a part number of a part, an unnecessary amount of changes and auditing is required to make the alterations necessary wherever the part appears.

7. The format of the routing and the sequence of operations from a rigid structure which may not be the most economical method of manufacture for some of the parts.

    The process sheet shown in the example is, in reality, another form of a bill of material system. Consequently the redundancy created is a roadblock to accuracy. Actually, better methods are known to combine process sheets and a bill of material. By listing the operation number on the bill of material much duplication of effort can be avoided. This method does not change the need for individual part drawings.

A word needs to be said about cost accounting. There is some fear shown by accounting personnel that simple parts requiring little manufacturing time will cost an inordinate amount to record time for each operation. There is, in fact, no need for each part to have each operation time recorded. Grouping of such simple parts will, in fact, not only decrease set-up time but decrease job cost accounting. It is also not necessary to stock parts unless the economies of scale make such stocking sensible. The *economical order quantity* calculation is a viable tool for determining the need. If all parts required for some time period, for instance, were issued by compatible groups, accounting for them in work-in-process by part number through completion does require some changes to manufacturing practices. It is quite feasible that, just as for machined parts, it will be found economical to produce large quantities of some parts. The capability for this flexibility only exists if piece part drawings exist.

Another tool in manufacturing which needs to be considered is called *group technology*. A definition follows:

Group technology   A method for manufacturing small-lot batches of parts with somewhat different geometries, features, materials, and/or sizes on small groups of specifically tooled machines located together, scheduled as a unit, resulting in mass production benefits.

In practice, fabricated parts that are similar in raw material and which may be sheared or burned from the same raw material meet these criteria. Thus, group technology could be a practical method to introduce mass production techniques into a fabrication shop. As much as a 70% reduction in set-up time has been documented within companies using this method. A job shop, or small lot manufacturer obtains the most benefits. Once again, piece part drawings are the prerequisite for this system.

FAMILY DRAWING PREPARATION

Many companies now utilize what may be called "format," or "family" drawings. In principle, the application saves a great deal of drafting time. Pump manufacturers in particular have a highly formalized means for preparing such drawings. Since pumps are a highly stylized product, each significant part may be designated using a name, or "noun code." The name also describes a particular part usually with a specific shape. Thus each of these parts may have a format drawing complete except for dimensions, notes, material, etc. This principle may now be expanded to encompass almost all kinds of parts, particularly those which can be described in a classification system by its shape made of permanent characteristics.

There are some disadvantages to this procedure, in that the final drawing will not be to scale, and it may be difficult to add or delete variable features. Also, in companies having a larger variety of shapes, a manual system using format drawings can become cumbersome and, indeed, lower the quality of the final drawing. The coming revolution, CAD/CAM, to make drawings on a cathode ray tube will, however, eliminate the problem, since family drawings can be stored in memory, called out using a classification system, altered to the specific part required, and then to scaling the drawing to the exact dimensions using a computer. We will discuss the principles of family drawings in another chapter.

COMPUTER-AIDED DRAFTING

It would seem as though all the past sections of this chapter have been needed to emphasize the disciplines required to fully utilize computer-aided drafting. It would be naive to believe that using a computer to make drawings is the ultimate solution to increasing productivity in a drafting department. On the other hand, it is also naive to believe a computer drafting system can be suddenly plunked down into a drafting room and be operated effectively. Two ways to approach computer-aided drafting are possible, or what might be considered a "push" approach, or a "pull" approach.

In the first or *push* method, a drafter would prepare the drawing required. In essence, the cathode ray tube represents a clean piece of paper onto which the drafter places the latest design. Let us take an example. Assume you are to draw a set of cross bracing for a building as shown in Figure 36. The layout will require you to scale in the column center lines, and the horizontal center line of the braces. The center plate

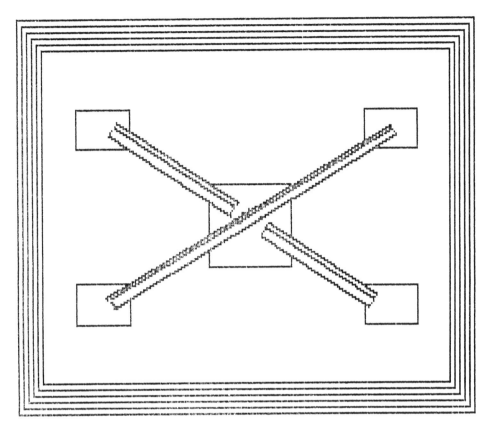

Figure 36 If the philosophy "we never make the same thing twice" is carried over into the means by which CAD/CAM drafting systems are implemented, the system will be highly inefficient.

must be scaled in, and then a particular angle, such as a $4 \times 4 \times 1/2$ inch must be drawn in, and then scaled not only to meet the face plates correctly and to their center, but also so the angles which meet at the center have no interference with one another.

The use of Smoley's tables may be appropriate to compute distances and angles during this stage. This method would seem to be quite time consuming and inefficient. As a matter of fact, it would not seem to be possible to draw as rapidly on a cathode ray tube as it might to simply take out a piece of paper and draw in the old way.

Suppose instead we use the "pull" system. What advantages will it have? We must first recognize the tools available to us:

A classification system

Family drawings

A cross-brace algorithm

Previous designs

A table of raw materials

With these tools, it is possible to first set up the cross-brace algorithm which has a generalized version of beam center-line and cross-brace heighth. By entering the dimensions required for these, it is then possible to interrogate previous designs to determine if any exist similar enough to the new requirement. If not, the pictures of the proper raw material, our $4 \times 4 \times 1/2$ inch angle, can be displayed on the center lines calculated and generated by the computer, and allowed to "cut" them so as to be properly aligned. Almost no line need be drawn "manually." There is a motto that the "pull" system implies.

It's Not How Fast You Draw Lines
It's How Fast You Find Lines Already Drawn

With this practical motto, and with the surrounding systematic tools in operation, computer-aided drafting will further enhance design, layout, and drafting productivity. The need for piece part drawings becomes readily apparent.

PURCHASED PARTS

We must also consider the documentation requirements of purchased parts. It would seem obvious that, since each vendor item is purchased as a unit, they would be documented individually and identified. Unfortunately, this is not always the case. For many years hardware items were given a description only. Part of the reason was the idea of engineers that were either too inexpensive to bother with, or they did not deserve the individual documentation needed to identify them. The mechanized bill of material as well as the increased cost of such items has changed all that. However, it has been replaced with totally inadequate documentation in many cases, as $1/4 \times 4$ inch hex-head bolt, or A. B. C. company bearing, vendor 4455A2. These descriptions do not allow a purchasing department to find multiple vendors, and obtain competitive bids. It is also not conducive to reuse as the next designer must look up vendor 4455A2 to see what it is. Again this possibly takes longer to do than document a new part.

Some companies use the vendor number as a part number. Some companies have no record of purchased parts except on the bill of material on which it is used. Each of these methods makes it difficult for someone else. Inspection has a difficult time receiving material. Service personnel have a difficult time replacing a part or shipping a reasonable alternate. When a vendor obsoletes a part sheer panic results. Hundreds of parts are excess in inventory. Retrieval is nonexistent. So-called standard parts catalogs have many tabulations of obsolete parts, parts which have been replaced, and parts which have never been used. In some extreme cases of over documentation, some parts are described based on standards of fit, finish, or tolerances dated 1957, for example, but no one is sure whether the parts received meet these specifications or, worse, whether they need to meet them.

It is necessary to create a document identified with, and identifying, a single item utilizing a singular unique identity or part number. Many of these documents can be simple typewritten A size drawings. Elaborate drawings are not necessary with pictures. Clear, concise, and carefully written descriptions are needed.

SUMMARY

With care and study, it will probably be found in the majority of manufacturing companies that the use of single piece part drawings, and their reuseability will increase productivity, and thus, decrease costs. With a good design retrieval system, success is almost wholly guaranteed. We have now discussed some of the reasons why preparation of piece part drawings is logical and economical. We can therefore close with some worthwhile questions:

1. If all parts do not require individual drawings, under what set of rules shall we be instructed to make those which are required? If the differences can be defined, can the two or more systems be made compatible?

2. If it is true that for full advantage to be taken in using piece part drawings, we must change present practices, have we the desire and incentive to so change or are we rooted too strongly to the past?

3. Will a company take the time and expend the money necessary to initiate a serious study to confirm the economies implied here?

4. Will a manufacturing concern recognize that an integrated system means all — not just part — of the requirements must be met to utilize such a system?

We hope this chapter has served to at least initiate questions if it has not served effectively in answering them.

DESIGN RETRIEVAL

5

POPULARITY

If a design retrieval system is indeed an important tool and factor in modernizing engineering systems and documentation flow, why aren't such systems more popular? Why do so comparatively few companies utilize more modern classification systems. An investigation of some possible reasons for this reluctance will do much not only to help to understand why such systems are so comparatively rare, but also to provide some insight into the reasons why all engineering documentation systems have not kept up with modern technology.

AMATEURS CLASSIFY

For many years companies have used various classification systems. As a matter of fact, few would be without them. However, classification systems and *retrieval* systems are different. The former puts designs into "pigeonholes." The latter finds them. If the pigeonholes are redundant, ill designed, unuseable for rapid retrieval, parochial in intent, and packed to overflowing, engineers, designers, and drafters won't use it.

The purpose of a classification system is to identify a design in such a manner that it can be easily retrieved. Retrieval is therefore the most important benefit of a classification system.

We have already discussed several classification methods including name coding and random coding. These two methods predominate in most manufacturing companies, even though they were designed many years ago and have since fallen into disfavor. Therefore, they have not been maintained and often do not reflect the latest product designs of the company. These systems can be pointed to as a reason why classification systems fail to work.

Let us look, then, at some of the reasons why modern classification systems have not been implemented as a generally accepted tool.

"SELLING" FROM THE BOTTOM UP"

Over the last 30 years, an unfortunate management policy has developed. Although unwritten, this policy essentially states, "no new system will be installed if the 'troops' don't like it." This attitude has, no doubt, helped maintain the status quo. How often today do you hear systems analysts, department managers, creative individuals say they have to "sell" their program. Perhaps this democratic approach has merits, but it is also a fine means for management to be less than interested in new programs. In addition, if engineers and designers feel a new system may put them out of work or reduce their ability to create, how can they find any incentive to be "sold" on a retrieval system. Too often management personnel have not only been uninvolved, but have not taken the time to investigate and understand such projects. By remaining neutral they often "kill" the proposal.

Selling from the bottom up may remain an essential part of our democratic process, but the "top" had better get involved at the beginning or they may find themselves with no bottom reporting to them and without a job themselves. Management by consensus has slowed progress. Conversely, many projects "sold" from the bottom up have been ill conceived and detrimental to the fortunes of the company simply because management has accepted the consensus without investigation and education. Both ends of the company would do better to sit down and educate each other. Thus, participation by management has been a significant missing link in implementing modern classification systems.

CAUSE AND EFFECT

In many companies, manufacturing personnel have given up attempting to show the engineering department how designs could be improved or simplified for more economical production. Certainly few definitive studies have been made in manufacturing to prove duplicate or closely similar parts exist. *Value analysis* studies usually emphasize a specific product, not a broad production study. With no "feedback" from manufacturing, there seems to be no "cause" to change, ergo, no "effect." As mentioned previously, the engineering departments of companies tend to live in more of a vacuum than any of us are aware. Many companies have their engineering departments in cities far away from the actual production center. Communication between the two forces are almost nonexistent. In one company, where a retrieval system was quite obviously necessary, the foreman of the fabrication shop had stacks of drawings. The back of each one had written instructions for the drawing next on the pile, including reference instructions for identical and similar parts. He said, "I'm always afraid to go duck hunting, because someone else takes my job who doesn't understand my 'system,' and it takes me weeks to straighten it out when I get back." These remarks and the atmosphere within which such a foreman works have not been communicated to engineering departments. Thus, a common question asked by engineers is, "why should we change"?

SHORT-TERM MANAGEMENT GOALS

Another phenomenon which has hurt the introduction of new systems into engineering as well as other departments has been the need of management to achieve "instant" success. Pressure for increased sales and increased profits have caused managers to look ahead to the next stockholders meeting, not to a more modern department. ment. An article in a magazine several years ago stated that a survey of several company board meetings showed 80% of all the subjects discussed concerned means to reduce taxes. Obviously, this concentration of activity left little time to consider the future.

Design retrieval systems take time to design and implement. It may also be one or two years before they begin to show savings which can be identified. It also takes a long time for the influence of a retrieval system to be felt on the majority of industrial engineers, shop management, tool designers, and the like. Thus, the implementation of a retrieval system is not a short range profit maker. Effort has thus been diverted from such projects to parochial, often ill advised short-range "fixes." Many companies have also found themselves carrying the burden of "caretaker" management, personnel who are not about to start a project that may fail and cause them early retirement. Thus, nonglamorous retrieval systems have suffered from neglect.

POOR MARKETING AND PUBLICITY

It is also unfortunate that those consulting firms who design and install classification systems are not as skilled in marketing their own product. Many years ago the head of a consulting firm specializing in design retrieval systems refused to show a client an example of his methodology. He was fearful the client would "steal" the idea and use it himself. Obviously, a consulting firm skilled in the art of classification has only one product to offer, the skill of classifying. Thus, there is a natural instinct to be careful not to disclose their technique. To a great degree, this problem has been alleviated. Yet, there is still a tendency by such firms to be close-mouthed.

In addition, the major consulting firms performing classification systems design originated in Europe, and were not aware of, nor skilled in, the marketing and publicity methods of the United States. The vast majority of firms utilizing classification and design retrieval systems are, today, in Europe, Africa, and the Union of Soviet Socialist Republic. The Soviet Union is the leading user of classification, coding, and design retrieval.

These consulting firms are also quite small, they have little capital to invest in promotion and rely, instead, on reference selling. This method can take years, since those companies who implement a system may not be sure of its value for a long time. On the other hand, if a classification system is effective, there is also a tendency for a company not to brag about it lest competitors also gain an equal advantage by implementing a like system.

In fairness to those consulting firms who implement such systems, their work is truly high quality, designed with great integrity. Rarely do the recipients of their experience voice dissatisfaction. Usually those systems which fail, fail as a result of the user company not pursuing the disciplines necessary for success.

DENIAL OF DUPLICATION

No designer or no drafter, likes to admit he has prepared the same design over and over again. One company in England was implementing a classification system. As the drawings were being scanned, a rather complex design was found twice, once drawn in 1937, once drawn in 1956. When placed over one another, they matched almost exactly. (These drawings were not made from a sepia or any other reproduction process.) Since both designs were made by the same person, it was obvious he was a consistent and meticulous person. These character attributes did not eliminate the fact, however, that he had prepared duplicate designs. Today it is not necessarily true that the same person will make duplicate designs, but in a design office where many designers and drafters work, duplicate parts are surely being prepared. Too many samples and surveys have proven this to be true to believe that there are many exceptions. Engineers, designers, and drafters simply do not wish to believe it.

Of far more importance than exact duplicates is the discovery of many designs which are closely similar. While some excuse might be made for this, the fact is that, had the drafter known of another design's existence, he could have used it without alteration. One has but to look at a design, any design, and note that as many as 80% of all the dimensions could be changed and not affect the function for which it was made. With this large latitude in the ability to make changes without hurting the design concept, similar parts are certainly quite likely to be changed to exact duplicates. Until engineers and designers really believe this concept, they will be reluctant to use avoidance of duplication as a reason for utilizing a retrieval system.

COLORLESS SYSTEM

The most dangerous part of a retrieval system is that, when it works, there is no "noise." One major electronics firm instituted a classification system and used it for 4 years. The whole industry fell on hard times and management went searching for places to cut overhead. The classification system stood out as an ideal place to "save money." The entire department was summarily dismissed. Fortunately, the general managers of the divisions within the company found out about this decision quickly before the personnel involved had packed their bags. The general managers pleaded with the corporate management to reactivate the system and offered to pay for it through their division budgets. This action was taken, and 8 years later the system is sound and healthy. Lack of publicity and awareness almost killed it.

It is true that a classification system has little glamour for the outsider. It is those who work with it who begin to realize the potentials for mass purchasing, better routings, similar tooling, increased design quality, consolidation of previous designs, and other attributes who get excited. The people who work in a classification system are not necessarily marketeers, however, and thus they do not tend to publicize their findings, nor do they get a chance to show their prowess or results to management. Thus "colorless" is the word. This, too, must be changed to allow a classification system to take its well-earned place as a "must" within an organization.

LITTLE COMPETITION

It is pretty difficult to tell a successful company it can improve profits with a classification system. Since it is a colorless system, management tends to think of better dealerships, better financing for branch warehouses, better and more modern machine tools, and other "things" as a means to promote success. Tangible changes have always been far more popular than intangibles, and a retrieval system is pretty intangible to most people.

Throughout the past 30 years, American companies, at least, have not felt the need for "tightening their belt" by reducing the luxury of having excessive raw materials and designs. Many companies have been proud to say they could sell all they could make. Expending already short engineering manpower on some freakish system as design retrieval when 40 customer designs are past due simply did not make sense. Thus the years in which design retrieval systems could have been utilized to prepare for real competition passed uneventfully. Most companies must play "catch up." Let us hope management won't ask for "instant" change. After all, it has taken 40 years for American companies to get to this spot. Why must we expect to change in a week or two? Only will a practical, long-range plan undo the years of neglect in conserving design variations.

We need to bring back design innovation similar to that shown by Henry Ford. His designers thought he was crazy when he so meticulously prepared specifications for the shipping crate for a battery. To their surprise, the crate became the floorboards of the car.

NUMERICALLY CONTROLLED TOOL SOLUTION

A program for implementing numerically controlled (NC) machine tools was embarked upon with great fanfare by many manufacturing companies several years ago. These tools were supposed to make it possible to manufacture inexpensively in lot sizes of 1 or 1000. The NC tool was supposed to be the solution to the high cost of job shops. This promise never came to fruition, except in rare cases. The difficulty in preparing good tapes for the machine, the proliferation of different tool path algorithms by the tool makers, and the lack of people capable of understanding both machine tools and computers contributed to management's disillusionment. The use of NC tools became limited to making higher quality parts faster when mass produced. Thus, the potential of making it incidental as to how many new designs were prepared failed. Of course, even had NC tools been successful in economically manufacturing parts in lots of one (which they will one day do with CAD/CAM), the costs of preparing all the other documentation, the differences between one part and another would have been too costly. Thus NC tools cannot and will not solve the retrieval problem. Only a retrieval system will do that.

Will design retrieval systems become more successful and popular?

For all the "wrong" reasons, design retrieval systems will become more popular and successful. Management of companies searching for economies in engineering and manufacturing are becoming more aware of the technique. Other reasons, however, are causing a surge of interest and investigation. Let us look at some of the more important ones.

SHORTAGE OF GOOD DESIGNERS

Engineers and designers have been in short supply for several years. Good innovative designers are even more rare. Many of the latter have not been utilized fully, since, as stated previously, engineering departments did not create many new systems or designs. Today competition makes innovators premium personnel. There is simply no reason to have such persons reinvent the wheel. Thus, it will be important to have at their disposal design entities which can be merged into newer designs. We must emphasize that even the most innovative design is simply putting together parts as with a Tinkertoy set. Thus, having the Tinkertoys at a designers elbow, readily retrievable, will improve the productivity of design.

LESS MASS PRODUCED PRODUCTS

Even if design innovations are not required, engineers today are finding customers increasingly require something other than standard. More and more design changes with slight alterations are being made. This is an ideal situation to utilize parts which have been manufactured previously, (and decrease lead time for the product).

"FAMILIES" OF SIMILAR PARTS

CAD/CAM systems are being cited as a means of increasing drafting productivity. These systems are rather expensive t-squares. Any time a line does not have to be drawn on a terminal, savings are made. This is why many companies are exploring the use of a "family" drawing representing several parts as entry into the graphics data base. A family drawing can be brought to the screen and modified quickly to make a new design. This will eventually provide continuous and profitable new parts for an engineering department.

   Be aware that finding similar parts or using family drawings to make new designs is not as onerous to a designer as the statement "eliminate duplicate designs." Creativity has been perpetuated.

CONSERVATION OF MATERIALS

It wasn't so long ago that the United States as a whole felt natural resources were unlimited. More exotic and slightly different steels, cast irons, titanium, and others were being specified for new designs. Today identical parts can be found, the only difference being the material specification. A study of these will no doubt find the designer used some of the new materials simply because they were there. To simplify raw material variety, reducing inventory, as well as restudying whether such materials are truly necessary will reduce design proliferation. A design retrieval system is an excellent tool to make such a study.

AN INTEGRATED CAD/CAM SYSTEM

Integration of engineering disciplines into a CAD/CAM system will be the largest impetus for increasing the popularity of retrieval systems. Such a system is the "glue," the "paste" which allows the merging and collection of data in a logical method. Many important facets of design can be stored by family, such as geometry, analysis programs, family drawings, master routings, tooling, and more. The ability to collect and disseminate this similar data will enhance CAD/CAM and make it a portent for the future.

   Figure 37 delineates three of the most popular results of families of parts and a

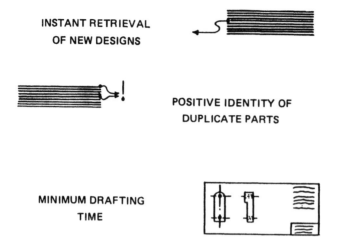

INSTANT RETRIEVAL
OF NEW DESIGNS

POSITIVE IDENTITY OF
DUPLICATE PARTS

MINIMUM DRAFTING
TIME

Figure 37 Three results of a classification system are instant retrieval, avoidance of redundancy, and less drafting time.

retrieval system. They are instant retrieval of new designs, positive knowledge of duplicate parts, and minimum drafting time.

TYPES OF CLASSIFICATION

We have already discussed some of the various classification systems. Let us look at two of the more modern and effective systems. They fall into two categories, a *monocode,* and a *polycode.* Figure 38 will define these terms.

Figure 39 is a graphic definition of a polycode system. Each of the important or key attributes of each part has been defined. Notice both a nut and bolt have been classified within the same system. The example, however, is not one of a good classification system simply because there are too many attributes classified for such a simple part. The attributes were prepared as though for design retrieval as opposed to utilization of the system where a polycode is extremely effective, manufacturing. Using polycodes to define manufacturing attributes for similar manufacturing and group technology makes a polycode an outstanding success. Thus, we will spend a bit more time discussing a monocode system for engineering design retrieval. After all, if the rules of a good retrieval system include the statement "all inclusive," we must be able to classify a part such as the one shown in Figure 40. A retrieval system utilizing a monocode technique usually divides the population of engineering designs into four major segments as shown in Figure 41. The examples we will use will emphasize that section called *proprietary* part design, or parts which are under the design control of the company. Figures 42–44 are pages taken from an example of classification prepared by the firm of Brisch, Birn and Partners, Inc., a consulting firm from Fort Lauderdale, Florida. Notice that, even with no words, the characteristic form and shape of a part can be used to follow a logical path to a family of parts. Thus, we can trace paths to family 201XX and family 253XX. The means by which these "paths" are made can be illustrated by two examples. Let us suppose we are looking for a part resembling a flat washer. Figure 45 shows the descriptors which might be used to find a family of parts meeting this requirement.

Figure 46 is another example. Here we see a "decision tree" used to find simple rectangular parts. When put all together in a logical network, the classification system might look like the page shown in Figure 47. It is here that we begin to realize how important families of parts are. Each block represents a group of families, and there are 100 potential groups. However, this utilizes only 3 digits of the code. An additional 2 digits of the code expands potential family drawings to 10,000. Now we can see how to plan a graphics system for CAD/CAM. Perhaps the first step is to develop a data base of families of parts. Purchased parts are classified in a similar manner. However, the function of the part is emphasized over the shape or form. Figure 48 is a catalog prepared as a result of a purchased part classification system. Notice how many categories of bearings have been segregated as a result of a concise sample of both type and frequency of bearing types within the company.

A word about the complementary uses of a monocode system and a polycode system. Figure 49 pictures a series of decision trees extracted from a monocode system. Each decision sequence leads to a family of parts. The descriptors used have been tailored to divide the parts into families with reasonably equal populations of parts. For example, all families of parts have not been classified with the specific descriptor "holes." It was not necessary to so describe the families. On the other hand, manufacturing may require such knowledge. Any and all parts with holes may require investigation to standardize on perishable tools, for example. Here is where a polycode can extract such data as shown in Figure 50.

This figure illustrates how polycode A might be used to select specific part characteristics from populations of parts. Utilizing a polycode attribute code with a design retrieval monocode is like preparing a cross matrix. Each method does not replace

CODING METHODS

MONOCODE
A MONOCODE IS AN INTEGRATED PROGRESSIVE CODING SYSTEM IN WHICH EACH DIGIT QUALIFIES THE INFORMATION CONTAINED IN THE PREVIOUS DIGIT. EACH DIGIT MAY CONTAIN MORE THAN ONE "DESCRIPTOR" OR CHARACTERISTIC.

POLYCODE
A POLYCODE SYSTEM IS A SET OF INDEPENDENT SEQUENTIAL CODES, EACH DIGIT OR SET OF DIGITS DESCRIBING INDIVIDUAL CHARACTERISTIC VARIATIONS. POLYCODES ARE OFTEN DESCRIBED AS "FEATURE" CODES.

Figure 38 Two predominant methods for classification of designs are defined as the use of a monocode or a polycode.

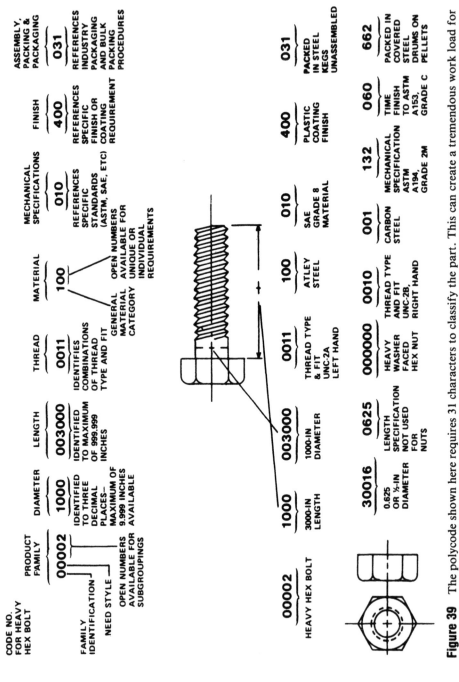

Figure 39 The polycode shown here requires 31 characters to classify the part. This can create a tremendous work load for classifying parts as well as maintenance. Care should be taken before designing any classification system to keep it from being overly long.

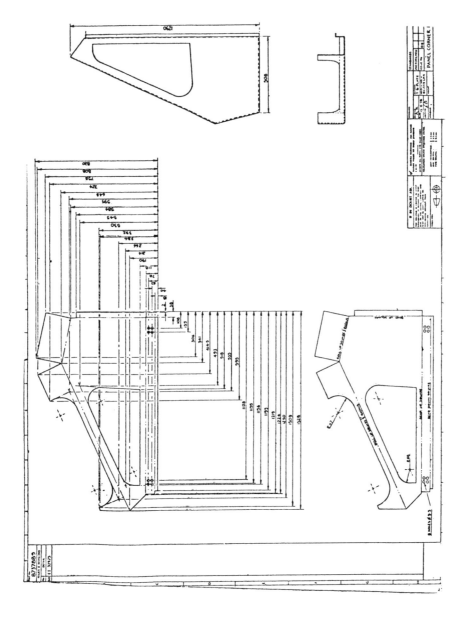

Figure 40 It is easy, when selecting a classification system, to overlook parts like this one which must also be classified.

87

Figure 41 Four basic categories of designs may be considered in manufacturing. These cover all segments of design.

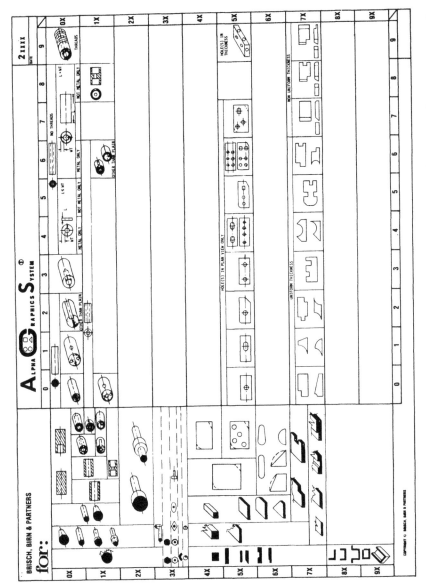

Figure 42 This "cybernet" code page makes retrieval comparatively simple using illustrations to augment the classification. (From Hyde WF, *Improving Productivity by Classification, Coding, and Data Base Standardization: The Key to Maximizing CAD/CAM and Group Technology*. Marcel Dekker, Inc., New York, 1981.)

89

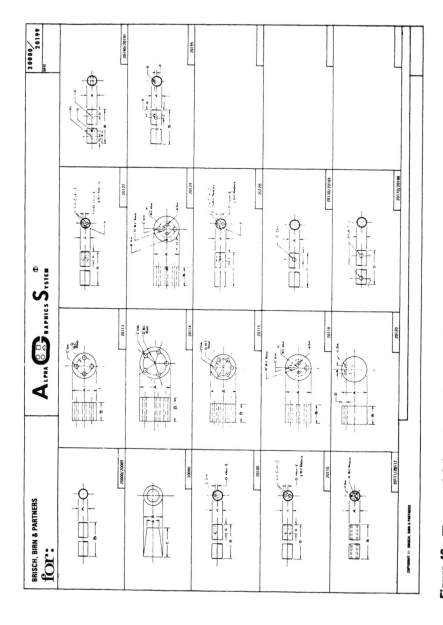

Figure 43 The second Cybernet page of two is used to determine the final two digits of a five-digit code. (From Hyde WF, *Improving Productivity by Classification, Coding, and Data Base Standardization: The Key to Maximizing CAD/CAM and Group Technology*. Marcel Dekker, Inc., New York, 1981.)

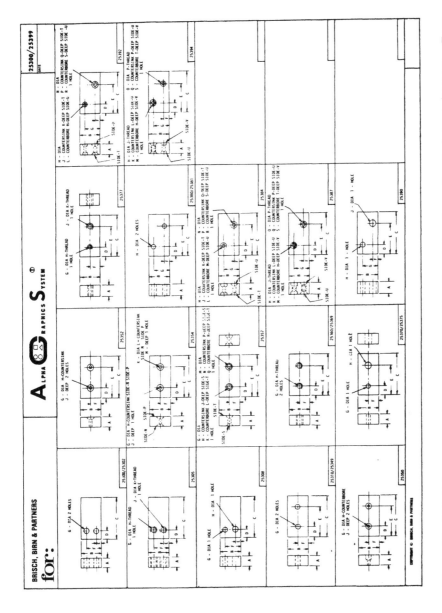

Figure 44 This Cybernet page shows alternate families of parts based on other design requirements. (From Hyde WF, *Improving Productivity by Classification, Coding, and Data Base Standardization: The Key to Maximizing CAD/CAM and Group Technology.* Marcel Dekker, Inc., New York, 1981.)

AXIAL
 ROUND
  SINGLE OD PORTION
   CENTER HOLE
    THROUGH
     THIN ONLY
      LENGTH EQUAL TO OR LESS THAN 1/4 THE OD

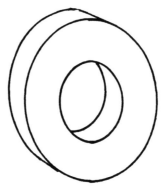

Figure 45   This typical summary of permanent characteristics of a design family clarifies the
sequence of decisions needed to find a part family.

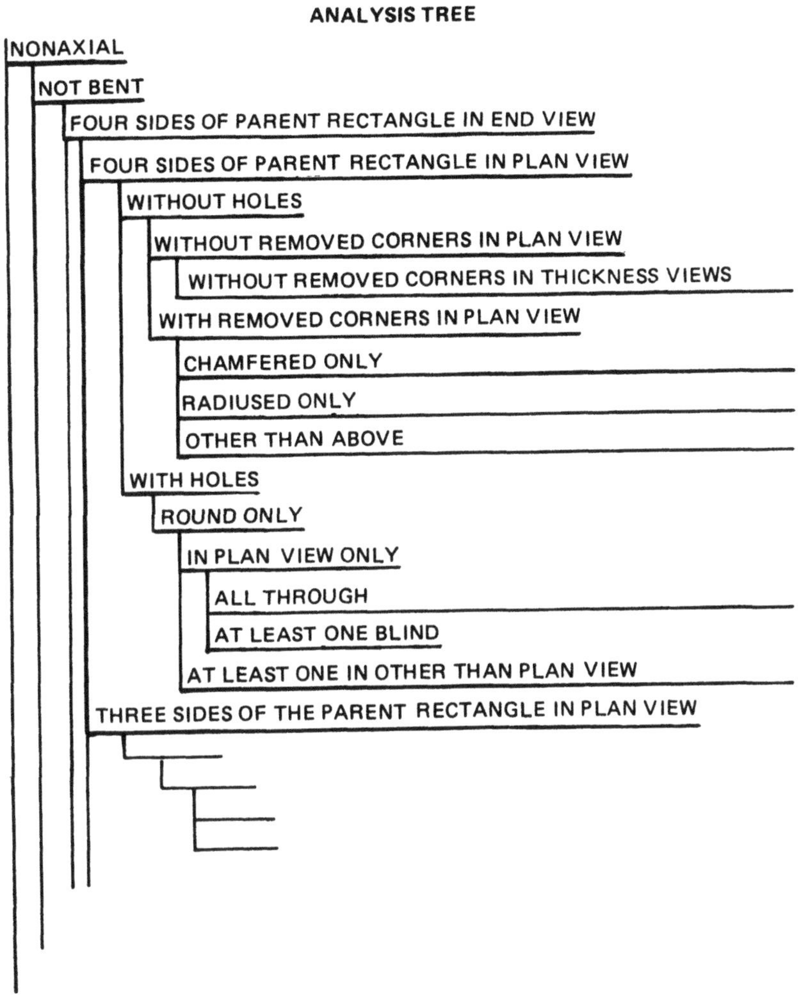

ANALYSIS TREE

```
NONAXIAL
  NOT BENT
    FOUR SIDES OF PARENT RECTANGLE IN END VIEW
      FOUR SIDES OF PARENT RECTANGLE IN PLAN VIEW
        WITHOUT HOLES
          WITHOUT REMOVED CORNERS IN PLAN VIEW
            WITHOUT REMOVED CORNERS IN THICKNESS VIEWS
          WITH REMOVED CORNERS IN PLAN VIEW
            CHAMFERED ONLY
            RADIUSED ONLY
            OTHER THAN ABOVE
        WITH HOLES
          ROUND ONLY
            IN PLAN VIEW ONLY
              ALL THROUGH
              AT LEAST ONE BLIND
            AT LEAST ONE IN OTHER THAN PLAN VIEW
      THREE SIDES OF THE PARENT RECTANGLE IN PLAN VIEW
```

Figure 46   An overall classification system using the monocode technique makes use of a "decision tree" in preparing the final system.

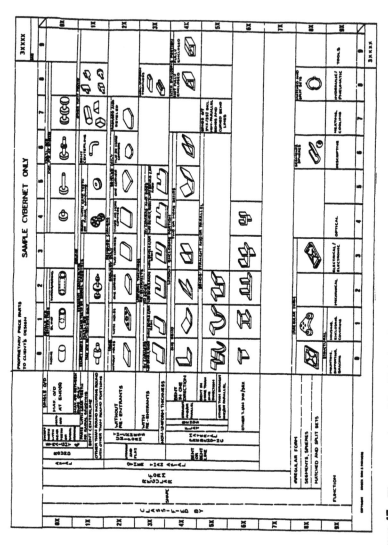

Figure 47 The segregation with which the planner is involved eventually divides the designs into 10,000 potential families of parts. Each box in this illustration represents a family, each of which may have a master routing, a family drawing, and all other manufacturing data needed. Planning can then be segmented into an orderly process. (Courtesy of Brisch Birn and Partners, Fort Lauderdale, Florida.)

94

COMMERCIAL PARTS
ALPHABETIC INDEX

20095	RIVETS	*BLIND
20070/3	RIVETS	*LARGE
20095	RIVETS	*POP
20095	RIVETS	*SELF-SETTING
20080/3	RIVETS	*SEMI-TUBULAR
20074/6	RIVETS	*SMALL
20070/9	RIVETS	*SOLID
20090	RIVETS	*SPLIT
20085/9	RIVETS	*TUBULAR
22282	RIVNUTS	
28628	ROCKER-LEVER SWITCHES	
25411/6	ROD-CLEVISES	
25400/8	ROD-ENDS	
27047	RODS	*INSULATING
26631	ROLL-FEEDS	
25308	ROLLER-SHAFTS	
25306	ROLL-SLEEVES	
25473	ROLLER-BEARING CARTRIDGE-UNITS	*FLANGED
25476/7	ROLLER-BEARING CARTRIDGE-UNITS	*CYLINDRICAL
25471/2	ROLLER-BEARING FLANGE-UNITS	
25450/6	ROLLER-BEARING PILLOW-BLOCKS	
25283/5	ROLLER BEARINGS	*AIRCRAFT-TYPE
25270/6	ROLLER BEARINGS	*AXIAL-THRUST
25200/17	ROLLER BEARINGS	*CYLINDRICAL-ROLLER
25240/3	ROLLER BEARINGS	*NEEDLE-ROLLER
25269	ROLLER BEARINGS	*SPHERANGULAR-ROLLER
25252/4	ROLLER BEARINGS	*SPHERICAL-ROLLER
25230/3	ROLLER BEARINGS	*STRAIGHT-ROLLER
25260/8	ROLLER BEARINGS	*TAPER-ROLLER
25234/5	ROLLER BEARINGS	*WOUND-ROLLER
26370/4	ROLLER CHAIN	
25307	ROLLS	
22727	ROPE	*WIRE
245 1/6	ROTARY-ACTUATORS	
24500	ROTARY-GEAR PUMPS	
24777	ROTARY SEAL-ASSEMBLIES	
28600/9	ROTARY SWITCHES	
28628	ROTARY-TOGGLE SWITCHES	
24502	ROTARY-VANE PUMPS	
26360	ROUND BELTS	
21030/49	ROUND-HEAD	*CAP SCREWS
20620/9	ROUND-HEAD-CROSS-RECESSED	*MACHINE SCREWS
20600/19	ROUND-HEAD-SLOTTED	*MACHINE SCREWS
21340/8	ROUND-HEAD SQUARE-NECK	*BOLTS
21365/21	ROUND-HEAD	*STOVE BOLTS-WITH-SQUARE-NUTS
20300/9	ROUND-HEAD	*WOOD SCREWS

Figure 48 Commercially available parts are classified predominantly by function. Often the wide variety of such parts are not recognized until after a home-made classification system is completed.

Decision Trees

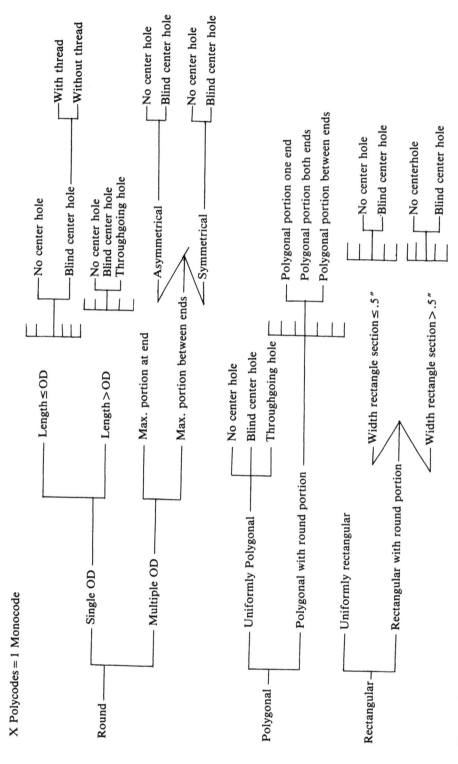

Figure 49 These decision trees illustrate the logical path through which a monocode system leads the interrogater. Logical and rapid retrieval results.

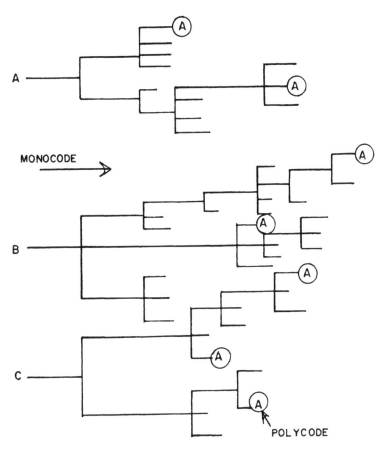

Figure 50 When separate features not included in a monocode are required, such attributes may be classified using a polycode.

the other. Rather they do, in fact, complement and reinforce each other synergistically. If we put all the pieces together, we can use the power of a design retrieval system. Figure 51 implies a data base designed on the basis of part families. Entry into the system is by interrogating the retrieval system until the proper part family is found. At the designer's disposal is not only a family drawing, but also a master routing, design tips, necessary polycodes, a family cutter center line for parts requiring numerically controlled machining, and a catalog to find specific parts. This is a powerful set of tools.

Let us look at an example to show how the system might operate. Figure 52 shows pictures of family drawings each with their respective family number. The drawings are generic in that no dimensions other than alpha characters are displayed. The geometry of the part as well as drawing data is stored in the computer. A computer program, tied to the family drawing is available to display entry areas for actual dimensions and redraw the specific drawing to scale in about thirty seconds. As we look at the flow in Figure 53, we begin with a new design requirement. Using the terminal, and following the design retrieval logic, we discover the family of parts similar to our new

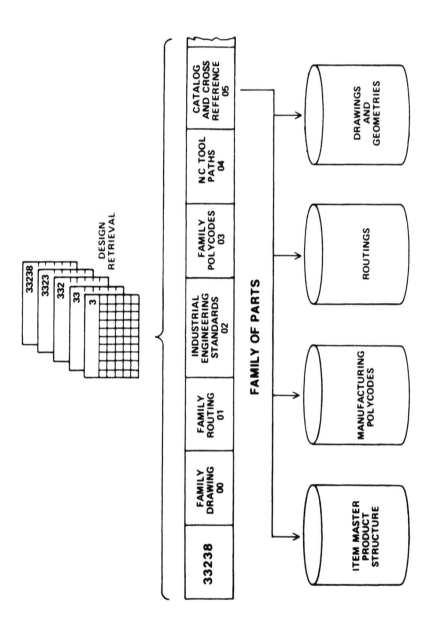

Figure 51 Here is shown a summary of the family data base reached through a classification system. Once in a family, specific parts and associated data may be obtained. The sequence of events allows an orderly storage of data in a logical array. (Courtesy of Control Data Corporation, Minneapolis, Minnesota.)

98

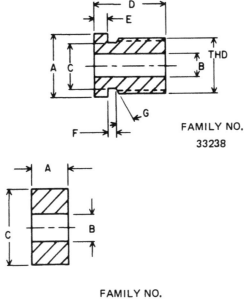

FAMILY NO.
33238

FAMILY NO.
30349

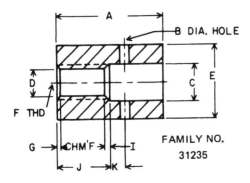

FAMILY NO.
31235

Figure 52 These three drawings represent families of similar parts utilizing shape and form. Each family has a number as 33238, 30349, etc. The drawing has generic alpha dimensions which can be replaced with actual dimensions when required. (Courtesy of Control Data Corporation, Minneapolis, Minnesota.)

design requirement is 30349. Viewing the family drawing confirms our logic, and we next place a catalog of parts for the family on the screen, if we find one we can use, after confirmation by viewing the actual drawing, we release it and we're through.

If we find none, we once again use the family drawing and its appropriate program, enter dimensional data only, and let the computer draw the new design. That's fast.

Classification systems are not designed casually. It takes professional skill to design one for a client. Figure 54 enumerates the three basic steps to a good system. Rather than "guess" about the segments required in a retrieval system, a sample of parts is usually taken from the file of past designs. This identifies the population. Next

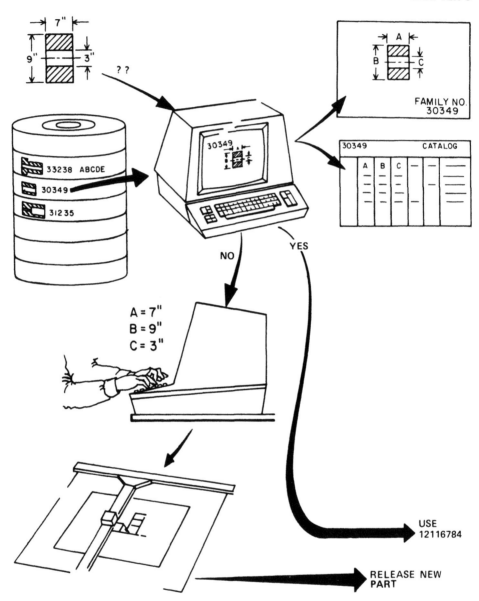

Figure 53 If a new design is required as shown in the left-hand corner, a search for the appropriate family drawing may find a design equivalent which can be used. If a design is not found, the family drawing can be placed on the screen of a CAD/CAM system and the generic dimensions replaced with the proper ones. This is a fast way to generate a new design. (Courtesy of Control Data Corporation, Minneapolis, Minnesota.)

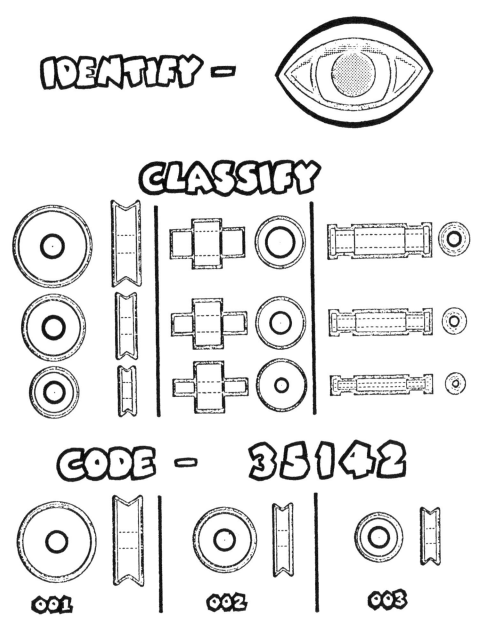

Figure 54 Three steps are required to implement a classification system, identification of the part to be classified, classification of the parts, and then only, a code applied to the designs.

the designs are "sorted" by their permanent features into representative families, and the distribution of this segregation carefully checked. Only at the end are the *code numbers* attached to the families of parts. This sequence assures adequate design coverage.

We all often tend to think of parts we wish to retrieve as round ones. Such parts are comparatively easy to classify. Figure 40 shows a more complex part.

Brisch, Birn, and Partners Inc. have investigated the frequency of part geometries throughout industry. Figure 55 shows the frequency of distribution through several major industries. Notice that 50% of all parts do not lend themselves to classification by shape but, rather, by function. In the same illustration, the second graph shows that, in selected companies, functional descriptions are even higher then the 50% average. This seems to show greater skill is necessary than we might have imagined for a practical retrieval system.

Figure 56 indicates another meaningful statistic in the art of classification. The first graph shows the distribution of parts other than round to be approximately 62%. This also indicates the need for more than average skill to concisely classify a majority of parts.

The second graph shows that, even in the electronics industry, use of round parts is only approximately 30%. Companies will do well to seek professional assistance when embarking down the path of implementing a retrieval system.

Besides the reasons we have listed for implementing a retrieval system, cost of new designs plays an important part also. Figure 57 shows the activities required to place a new design into production. Obviously this is only a skeleton of the many detailed jobs surrounding each of them. Many studies have been made to determine the cost of issuing a new design. Summarizing these activities, Figure 58 accentuates the steps. First, the drawing is made, a bill of material must be prepared, a make or buy decision is discussed, all sorts of conversations may occur between an industrial engineer and the designer, the part must be routed and costed, and the part may be made as one of a kind instead of with a batch of parts. Finally, new tooling may be manufactured, and the part must then be placed in inventory.

As one might suspect, releasing a new design can cost from $200 to $2200 before the first "chip" has been cut. One manufacturer made a survey of costs which are shown in the table in Figure 59. From left to right in the table are various parts from most complex to simple. The first part, for example, is a forging, the last a fabricated part. The engineering, manufacturing and data processing cost totals vary from $391 to $6420. The survey was made in 1978. Notice that drafting hours are not the major ingredient in any but the simplest design.

We cannot close this section without calling attention to a more comical side of design. Classification systems allow the user to study segments of parts for inconsistencies and other characteristics. The following illustrations are in recognition of the drafter who never makes errors but, rather, "abberations." These parts are, incidentally, being manufactured currently.

You can tell the drafter had had many unfortunate experiences with industrial engineers when you look at the designs in Figure 60. Notice how the diameter of the bore on one part is noted, "to fit transmission." This allows any red-blooded machinist to make the part any way they want without criticizing the designer. The other part, with tapered bore clearly shows the diameter to be 3.500–3.501 in. with a taper 1-1/4 in. per foot. From which end of the part should the taper be started?

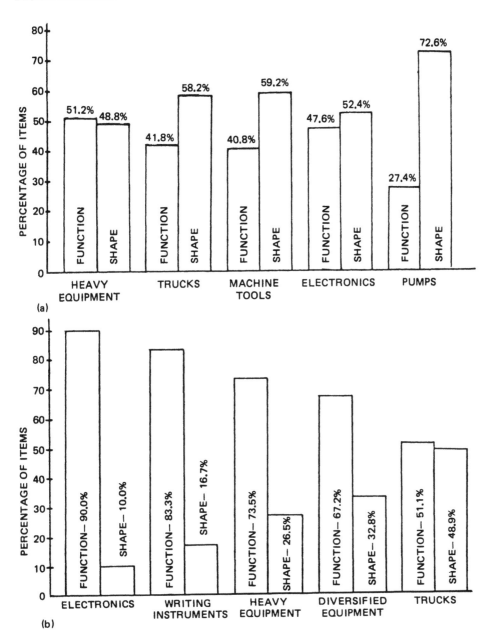

Figure 55 Studies of a variety of manufacturing firms show the diversification of designs by shape and function, including extremes for some. (Courtesy of Brisch Birn and Partners, Fort Lauderdale, Florida.)

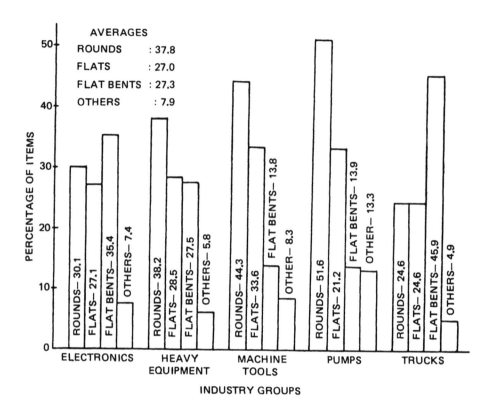

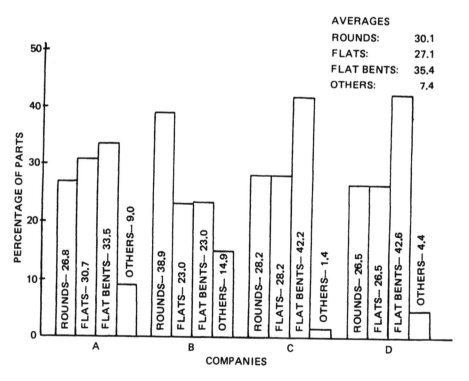

Figure 56 Within a variety of industries, round parts are only a portion of the total designs. Thus a classification system must accommodate a variety of shapes. (Courtesy of Brisch Birn and Partners, Fort Lauderdale, Florida.)

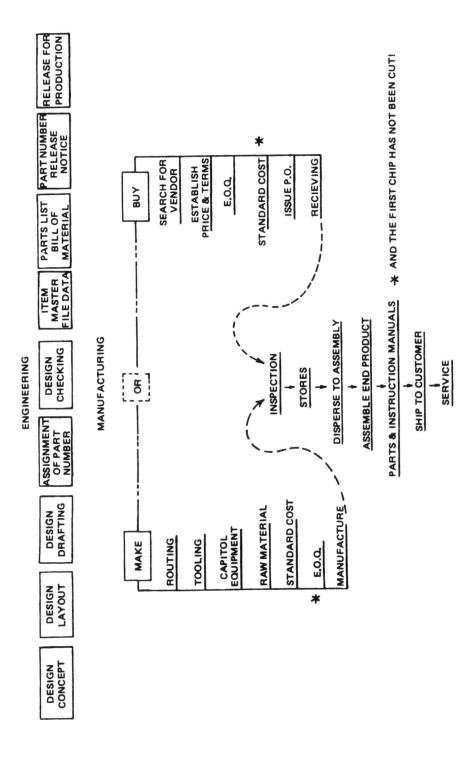

Figure 57 Initiation of a new design causes at least 24 activities before it becomes a physical entity. Thus the cost for initiating a new design is not trivial.

WHAT'S IN A PART NUMBER?

Figure 58 When a new part number is assigned to a design, it triggers a host of activities besides making a drawing. A bill of material, make or buy decisions, questions from manufacturing, routing and costing the part, manufacturing the part one at a time instead of mass producing, new tooling and inventory are all initiated.

PART NUMBERS	713714	514683	405822	405859	405842	405815	713713	405821	405896	406208	520522	TOTAL HOURS	AVERAGE HOURS	AVERAGE COST
ENGINEERING														
DESIGN HR	80	10	8	18	10	6	48	3	6	6	8	203	18.45	$322.95
DRAFTING HR	40	5	4	5	6	3	16	3	3	4	8	97.	8.82	$154.32
SUPPORT HR	32	2	1	3	1	.5	32	.5	1	1	4	78.	7.09	$124.09
DOCUMENTATION HR	.1	.1	.1	.1	.1	.1	.1	.1	.1	.1	.1	1.1	.1	$ 1.75
														$603.11
MANUFACTURING														
ROUTING HR	2.1	.5	.55	.75	.4	.5	.8	.4	.4	.5	.1	7.	.64	$ 11.14
TOOL DESIGN HR	4	2	2.5	.5	2						1.5	12.5	1.14	$ 19.89
TOOL MFG. HR	208											208.	18.91	$330.91
INV. CONTROL HR	.5	.5	.5	.5	.5	.4	.4	.4	.4	.4	.5	5.	.45	$ 7.95
														$369.89
TOTAL HOURS	366.7	20.1	16.65	27.85	20.0	10.5	97.3	7.4	10.9	12.0	22.2	611.6		
DATA PROCESSING (DOLLARS)	2.75	2.75	2.75	2.75	2.75	2.75	2.75	2.75	2.75	2.75	2.75			$ 2.75
TOTAL (DOLLARS)	6420.00	354.50	294.12	490.12	352.75	186.50	1705.50	132.25	193.50	212.75	391.25			$975.75

NOTE: BASED ON AVERAGE HOURLY RATE OF $17.50/HR

Figure 59 Analysis of part costs with various complexities is shown here. Part number 713714 is a complex forging, while 520522 is a simple fabricated part. Costs vary from $6420 to $391 to prepare a part for manufacturing. If using $400 as an average, each new design requires this amount to prepare documentation before the first "chip" is cut.

107

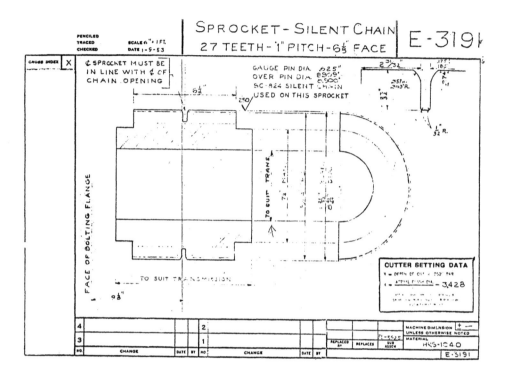

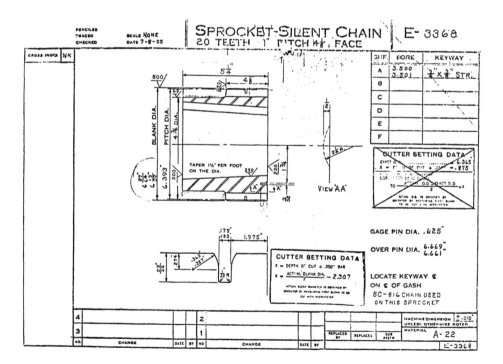

Figure 60   The tapered bore on this part starts from one end. Which one?

It is probably better not to commit yourself at all. The drafter who drew the drawing in Figure 61 decided to leave the hole diameters off entirely. Thus he cannot be criticized for dimensioning them incorrectly.

It is apparent that the drafter who drew the part in Figure 62 has had 20 or 30 years of a tiring experience. He was simply tired of dimensioning using 1/4, 1/2, or 1/16 in., so he began to dimension in thirds.

With modern engineering buidings, today's engineers can't throw their mistakes out of the window. Some engineer had a surplus piece of shim stock and didn't know how to get rid of it. In a flash of brilliance it was designed into a part. The part was sawed into two sections, the shim inserted between them, and the two sections rewelded together, as shown in Figure 63.

Most engineers are fearful a design will fail in the field. One engineer decided to use an angle bar in a design as pictured in Figure 64, and began writing its size down as, 1-1/2 × 1-1/2 × __ in. Then he stopped, and said to himself, "I'd better be careful." He finished off the dimensions with the leg thickness of 1-1/4 inch. You can be sure this part did not fail in operation.

Last, but not least, most drafters are proud of their "spatial recognition" capability. This, of course, means the skill to picture a part in one's mind, rotating it in any direction and visualizing it. Such was the person who designed the part in Figure 65.

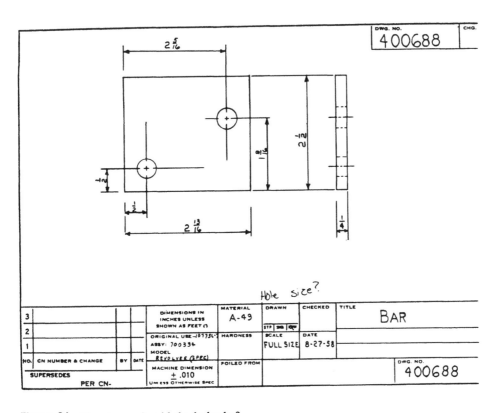

Figure 61 How large should the holes be?

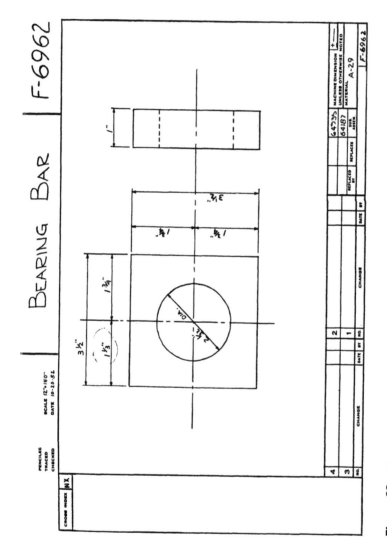

Figure 62 If you are tired of using 1/16, 1/8, or 1/4 in., how about 1/3 in.?

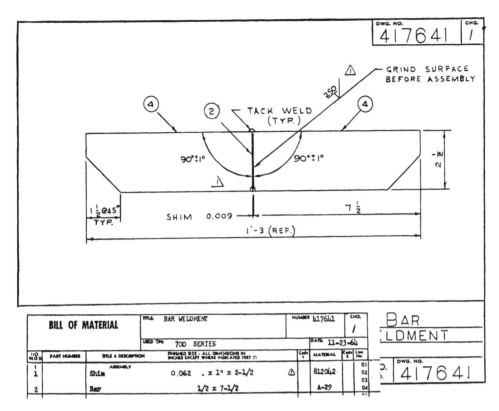

Figure 63 If you have a spare part you don't want, cut a design in two, put it inside, and weld it shut.

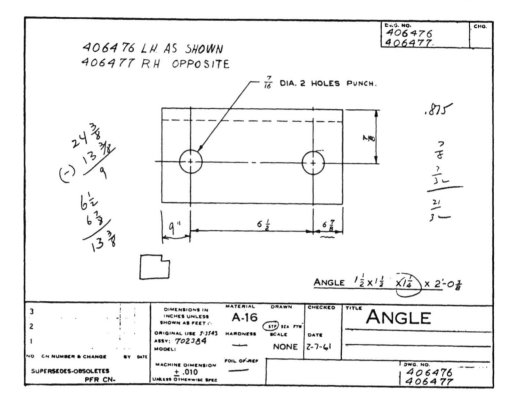

Figure 64 If you are worried about the factor of safety of a design you can always increase the dimension of the leg of angle at the last minute.

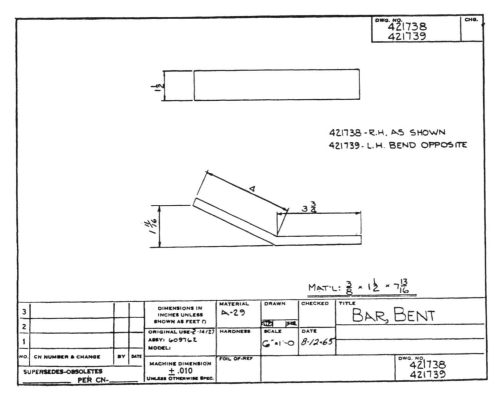

Figure 65 This design and the instructions are a test of your "spatial recognition."

CLASSIFICATION IMPLEMENTATION

The following major segments of activities are provided as a skeleton of a possible implementation plan for a classification system. To document fully these activities, a program evaluation and review technique (PERT) chart or an informal activity chart should be prepared, assigning responsibilities for action, cost to implement, savings possible, and time schedule for completion. If this activity chart is then approved, including manpower and time, the plan should be successful. The implementation activities have been divided into phases for simplicity and to avoid attempting to perform too many activities at one time.

Phase One
   Education program for engineers and drafters
   Microfilm-microfiche procedure
   Retrieval policy for new designs
   Retrieval rules for engineering
   Item master, bill of material integration
   Catalog update procedure
   Glossary of terms and definitions
   Alphabetical index
   Part number control and issue
   Multiplant rules to use system
      Retrieval
      Responsibility for classifying
      Cataloging maintenance
   Savings evaluation system
   Priority for adding new designs
      All new drawings
      All item master parts with usage
      All other parts on item master
      Parts in any old retrieval system

Phase Two
   Definition of preferred parts with rules
   Education program for all division personnel
   Policies, standards, and procedures for family drawings
   Producibility tips by family
   Part families for design analysis
   Family routing definition
   Part standardization by family
   Definition and use of group technology
   CAD/CAM system integration

PHASE ONE

These activities require some detailed explanation to fully describe them and gain insight into their degree of importance, and manpower or time required for each one. The following series of descriptions are meant to achieve this goal. It should also be noted that the implementation has been divided into two phases. The first phase is primarily within the engineering department, while the second phase reaches out and includes several departments other than engineering.

EDUCATION PROGRAM FOR ENGINEERING

There is a primary need to explain to all engineers and drafters how the classification system works. It is not necessary yet to completely explain all the important, but peripheral benefits of the system. The following outline is recommended for use in preparing visual aids and other material for this training:

Statement of policy by management
> Definition of classification
> Basic classes of parts
> Code page explanation
> Principals of a "mono" code
> Statistics of parts classified
> Scope of the system
> Example of retrieval and code pages
> Cost for issuing new designs
> Use of catalog
> Use of microfilm and microfiche
> Integration with item master
> Catalog update procedure
> Policies and procedures for retrieval
> Future uses and economies
> Method to report and record retrievals
> Savings documentation form
> Organization of variety management group

It is recommended that sample retrievals used in the presentation take the form of a "test," thereby allowing each person "hands on" experience. The presentation should be held for no more than 15 or 20 people at one time. It would also be highly advantageous to video tape the presentation for future education of new personnel. The visual aids combined with a short explanation of each one may be used as a handout to each person at the end of the presentation.

The training session should be followed by personal assistance for each engineer and drafter from a member of the classification and design retrieval section.

MICROFILM-MICROFICHE PROCEDURE

There has been considerable discussion surrounding the use of microfilm or micro-fiche. In either case, it is mandatory to the success of the system to have rapid and easy reference to drawings. If microfilm is used, at least 6 copies of all drawings in the classification system need to be prepared and sorted by family number. Later, as more departments use the system, additional sets of microfilms will be required. There are several drawbacks to using microfilm. External films can be mishandled, and microfilms lost. Usually, a set of microfilms need to be studied, and they can be misfiled after use. The sheer volume of parts makes such a working file bulky and clumsy to use. Continuous up-date and maintenance for 6 to 10 files can, again, create a nuisance in misfiling and general maintenance.

Use of the "paper" file of drawings is no better, since only one file is available. Thus, the searcher must rely on the ability of the classification section for all retrieval. It is not economical to consider adding more "paper" files for external plants. Thus, using the file folders of drawings does not appear practical.

Microfiche, on the other hand, seems both practical and economical. If completely packed, a microfiche will hold between 40 and 50 drawings. Thus, 750 microfiche will hold 30,000 classified drawings. However, each microfiche should probably be designed to hold no more than 3 to 6 "families" of parts with allowance for expansion. Micro-fiche can be rapidly updated, and economically copied, (13 cents per copy).

The following requirements and cost are involved in implementing a microfiche system:

Camera, cost	$ 4,000
Jacket loader	1,900
Duplicator	3,300
Microfiche jackets	225
Film processing	90
Duplicate jackets	975
Total	$10,490

With about 5% exception, the small drawings prepared for file folders will be clear enough for converting to microfiche. This operation can be performed at the rate of 1000 drawings per hour. This should take 100 hours, including assembly of the microfiche "chips" into jackets for a cost of $600, (using $6/hour as cost).

With this procedure, expansion of the classification system as well as expanded use by others of the system is quite practical, rapid, and economical.

For those parts in the file folders which have poor reproducible drawings, a brief scan of each folder can be made to see which ones should be remade. Even if this is not done at the start, they can be made a part of the microfiche, and updated at leisure in the future. It should be remembered that a copy of these drawings will be used to examine the design, not make the part.

It should not be overlooked that microfiche viewers, with the ability to make copies from them are required also. The same type of equipment would be needed should microfilm be used.

RETRIEVAL POLICY FOR NEW DESIGNS

Several approaches have been taken by other companies using a retrieval system. Usually a company starts with a "benevolent" approach, asking each designer and drafter to use the system, and trying to gain his cooperation. This method is time consuming and usually unsuccessful. However, it is a source of continuous aggravation to a designer to be given more "red tape" if the system cannot be used rapidly. If the catalog and microfiche are implemented correctly, this failure of the system should not occur. It is strongly recommended that a "mandatory" retrieval system be implemented.

The retrieval policy should be written and approved by the general manager so that all personnel involved are aware of the requirement. This policy should be publicized, and provisions for following it clearly stated.

RETRIEVAL RULES FOR ENGINEERING

Once a policy has been established whereby all personnel involved in preparing designs and drawings are included in the retrieval process, rules for when it should be used should be established. The best results occur when designs are searched for during the layout stage of the design. It is therefore strongly recommended that those who make layouts receive additional training on the classification system, including assignment of a person from the classification section to each person who makes layouts during the learning cycle. It would be quite useful to assign each person who makes layouts to the classification center for a period of time until he or she gains a perspective of the system, and increases their respect of, and confidence in, the use of classification.

ITEM MASTER, BILL OF MATERIAL INTEGRATION

A new system works only as well as it can be easily integrated into other active procedures. The item master is the main engineering data base, and the classification and coding family and sector number for each new design should be added to a data base. An on-line input system in engineering should provide for this additional data increment. In addition, the present flow of data should be augmented to accommodate both the microfiche system, the addition of the class code, and the up-date of the catalog system for classification.

The following ideal flow of data will serve the engineering department efficiently:

1. When drawings are completed and reviewed, the designer or drafter should receive a part number from the classification section.
2. All item master data should be entered on line by the designer or drafter.
3. Drawings should be sent to an engineering services section.

4. The drawings should be sent to be microfilmed.

5. As the item master is prepared, the computer should automatically generate a punched card for the microfilm, which would then include the class code and sector number.

6. Engineering services should next pull the drawings returned from microfilming, and sort them by family and sector number.

7. The drawings thus sorted should next be sent through the microfiche process and the resulting "chips" of drawings inserted into the microfiche jackets.

8. The updated microfiche may then be reproduced in the quantity necessary.

9. As the item master is updated on-line, the critical dimensions needed for the classification catalog should also be entered to update the classification catalog.

CATALOG UPDATE PROCEDURE

If any present cataloging system is a batch-type system. It should be rapidly converted to an on-line inquiry and update system as rapidly as possible. Basic *menus* for this activity should be designed. As mentioned previously, this maintenance procedure should become a natural part of data entry for the item master and bill-of-material system. While, for a time, the two systems may have separate data bases, the two should also be merged into a unified set of data elements as soon as possible.

GLOSSARY OF TERMS

A glossary of terms, which defines specific *descriptors* used to retrieve designs, is almost complete. The glossary should be completed, and then converted into a small booklet, shirt pocket size, which can be made readily available to all engineers, designers, and drafters. This directory will be of great assistance to those who will be using the system for retrieval.

ALPHABETICAL INDEX

The alphabetical index is needed as another short cut for entering the retrieval system. While not as exact as the step-by-step inquiry, it can be rapidly used in some cases. It might be feasible to make this index a portion of the booklet of glossary of terms.

PART NUMBER CONTROL AND ISSUE

Today it may be relatively easy to obtain a new part number for a design. To control issuance of redundant designs, it is strongly recommended that the classification and coding section be responsible for issuing part numbers. It is not the intent of this recommmendation to create unnecessary red tape, but, rather, to assure a constant review of new designs to promote standardization and to eliminate redundancy. If it is recognized that issuing a new part number implies a cost of $1600 for each new design, the time or effort to obtain a new part number is insignificant by comparison.

If a new design is drawn and released and *then* examined by persons in the classification section, it will be a continued irritation to the person creating the design by "second guessing." It is more logical, and will create better personnel relations to have a cooperative effort in issuing new designs.

MULTIPLANT RULES FOR THE SYSTEM

Retrieval of old designs is an important part of the classification system. It is unfair to believe that remote plants should rely on the telephone to find such designs. Consequently, a person skilled in classification and design retrieval is required in each remote facility which creates designs. In like manner, it is this person who should be responsible for issuing new part numbers; this should not be the responsibility of the central classification center. However, the central classification section must have the responsibility of monitoring all new designs prepared by external plants, should, on rare occasions, a duplicate be made.

Designs prepared by external plants must, of course, be entered into the microfilm and microfiche systems as rapidly as possible. In addition, the update of the catalog is essential to obtain the full benefit of the system. Thus, catalog data sent to the item master system is important since this is where all engineering data meets in the present system.

SAVINGS EVALUATION SYSTEM

The most "dangerous" part of a good classification system is that, when it operates correctly, it does not automatically prove its value. Many companies utilizing a classifying system have, when searching for reduction in overhead, suggested elimination of classification systems. These companies are usually those who have established a classification system on a casual basis with no formal methods to ascertain the value of the system. It is not wise or necessary to be so casual. Thus, a rigorous means for determining cost and savings for the classification system needs to be devised.

Several companies have been successful in reporting savings of a classification system using a monthly report such as the one that follows:

1. Class of parts (purchased, designed, etc.)

2. Number of inquiries by class

3. "Hits," or found to exist, number and percentage of inquiries by plant

4. Duplicates eliminated

5. Total "find" rate

6. Total retrievals as a percent of total new designs issued

7. Consolidated "find" rate for all plants

8. Savings based on the find rate and complexity of the part (using a predetermined set of values)

In the yearly review of the system, the following data can be summarized to evaluate the system:

Classification section personnel costs

Drawing, reproduction, and microfiche costs

Travel, training, etc., costs

Gross savings

Net savings

A daily inquiry "log" may be maintained by the classification section to establish the reports suggested above. The data on such a log should include:

Inquiry number

Date

Name of inquirer

Family number interrogated

Part name

Comments

Part number found

Reference part number found

PRIORITY FOR ADDING NEW DESIGNS

When the original classification effort is completed, a group of parts chosen at random have been classified. During the design period, many new drawings have been prepared which have not been entered into the system. These designs should be entered into the classification system as quickly as possible, since they reflect the latest in design requirements. At the same time these designs are entered into the system, provision should be made to change the flow of drawing release to accommodate all current new designs so the system becomes current.

Since the drawings used were taken from the current item master data base at random, many other designs remain in the item master which have not been classified. However, it would seem important to evaluate by priority which of these, and in what sequence, they should be added to the system. Two methods can be used to perform this evaluation.

If the item master data base is searched to determine which parts not yet classified are presently in inventory, in manufacturing, or in spare parts inventory, those which are should be next added to the classification system. Another method which will locate parts with usage but not necessarily in inventory is to generate an ABC inventory analysis which generates total usage of all parts based on a yearly usage. This report, matched against the item master will also determine current and active parts which should be included in the classification system.

It has been voiced that "old" parts, meaning those designed five or ten years ago are unimportant. It should be emphasized that a design, no matter what its age is legitimate regardless of its age. Thus, retrieval and inclusion of designs cannot be evaluated by the age of a design.

Once the "current" parts on the item master have been entered into the classification system, another evaluation should be made to determine whether all other parts in the item master should be added to the classification system. A sample of these remaining parts will help make such a decision.

It is possible that many parts have been classified by some other means. Included in the new classification, provision should be made to enter all these designs into the new system.

PHASE TWO

The activities in phase two represent an expansion in the activities of the classification section and a corresponding increase in the value of the system, since departments other than engineering are involved. It is recommended that a task force be designated to aid in implementing the activities required. Not only must the activity be more thoroughly defined, but manpower and responsibility for action must be designated. Several of these tasks are continuous while others are programs requiring superimposing duties on people already busy in their regular work. Thus, without a plan of action and continuity of effort, these important and money-saving projects will not be completed.

DEFINITION OF PREFERRED PARTS WITH RULES

With the completion of the classification project design, and with the addition of all old and new designs, a repository of drawings has now been cataloged in a logical manner. However, many of these designs will be found to be closely similar, of expensive material, or have some other discrepancy which should not be perpetuated. Each family of parts will have its own unique problems.

Each family of parts needs an examination to set standards for the elimination of those least desirable. By elimination is not meant discarding the design, but to mark those which should be used as preferential. Later we will describe the need for part standardization by family. This activity could be performed in parallel to this task. However, it is important to distinguish between parts which are recommended for reuse and those which are not, so no additional activity should slow this process.

A temporary task force of designers, manufacturing engineers, and drafters should assist in setting the rules for distinguishing preferred parts from nonpreferred even, perhaps, setting up a procedure to pass file folders of parts to this task force in an orderly manner to set individual rules for each family. This method would not only utilize the knowledge and capabilities of such people, but also help them gain a better insight into the system.

EDUCATION PROGRAM FOR THE DIVISION

The original training program for the engineering department should be expanded to include training and education on the other benefits of classification and design retrieval. All the activities described in this report should be covered, with special emphasis on the benefits of these external to engineering programs. Some specific subjects which should be covered are as follows:

1. Preferred parts
2. Family drawings
3. On-line retrieval inquiry
4. Family routings
5. Producibility tip data
6. Group technology
7. NC tool data base
8. Design analysis data base
9. Part standardization
10. CAD/CAM integration
11. Multiplant operation
12. Tooling
13. Spare part organization
14. Purchased part classification
15. Part standardization
16. Cost and savings
17. Quality control

POLICIES, STANDARDS, AND PROCEDURES FOR FAMILY DRAWINGS

The most important aspect for a successful CAD/CAM project may well be the use and treatment of family drawings. Since, by design, all parts in a family of parts defined by form or shape are similar dimensionally, a data base of family drawings becomes the means for creating new drawings on an exception basis. The family drawing for a family should reflect the best standards to manufacture the part. This also means consideration should be given on where and how many dimensions and views are required. Standardizing location of dimensions, including the possible need of different methods for those parts which will be manufactured on numerically controlled machine tools will become important.

In examining family drawings recognition of *layers* in a CAD/CAM system will further make family drawings important. It should be recognized that sheet metal parts which are complex when formed or shaped can be shown one step at a time including a complete flat layout on various layers of the CAD/CAM system. Recognition of similarities of parts in a family allows more time to be spent in preparing family drawings, since any one of them will cover several parts.

Several different philosophies will need to be examined when addressing the family drawings, since many part families may require more than one family drawing, and parts classified by function as opposed to shape may require different rules and procedures. Also it should be noted that a family drawing may be created as a two-dimensional (2-D) model, but a computer program in CAD/CAM will prompt the user to fill in dimensions needed for a new design in the family and create as a result a *wire-frame* three-dimensional (3-D) drawing. Thus, the interrelationship of family drawings, engineering standards, and CAD/CAM is an extremely important consideration, and must be documented in a series of policies, standards, and procedures.

PRODUCIBILITY TIPS BY FAMILY

The use of producibility tips when designing new parts can become a most important area for savings in the overall project. Producibility tips are a reflection of how industrial or manufacturing engineering sees the part as the manufacturing process is prepared. Methods for dimensioning including the penalty of different tolerance requirements may be documented. The method to machine a part so that it will perform as the engineer or designer wishes can be shown. For example, it is not uncommon for a designer to machine a round part into a hex head should the part need to be turned by a wrench. It is much more economical to machine two parallel faces to perform the same objective. This type of alternate method becomes a portion of the data base for family drawings and a CAD/CAM system, and is closely tied to the part standardization effort.

Manufacturing engineering should be assigned the responsibility for working with the classification section to develop producibility tips. Since this is a long-range project, priorities must be set to examine part families based on complexity and/or quantity of parts in the family.

Much additional experience could be gained in the preparation of producibility tips if foremen of the machine shop and fabrication were involved in the study. In addition, many persons who make the actual layouts, and the templates in fabrication can contribute to the project.

PART FAMILIES FOR DESIGN ANALYSIS

For many years design analysis of parts and assemblies was conducted more or less in isolation. For example, gear design analysis produced a wealth of information about a gear which resulted in such data as tooth form, cutting data, cutter data, and general dimensions of the gear. This data was then laboriously transcribed from computer output sheets to the drawing for the gear. With a CAD/CAM system, the integration of this data into a family of parts will allow the gear and associated technical data to be automatically transcribed onto a drawing without the aid of a drafter.

Not only are several types of design analysis programs presently conducive to this form of combination, but the future of design analysis will create more possibilities. Thus, the most economical way to consider combining design analysis with drafting is to consider families of parts. Since a classification system creates the families, it seems only logical to define those families of parts presently parallel to the design effort. Then the data base for CAD/CAM will be designed such that families of parts will have the necessary geometry and programs associated with the family to make the transcription from design analysis to drafting automatic.

FAMILY ROUTING DEFINITION

A classification system which is based on permanent features of a part such as form, shape, or function is also compatible to segregating parts which are machined over the same work centers. It is true that such a classification system does not identify such specific attributes as milling, grinding, slotting, boring, drilling, etc., across all families. However, it does partition parts which can be routed over a series of identical work centers. Thus, if a family of parts is examined, an ideal or most practical series of work centers for the group of parts can be identified. In addition, similarity of parts indicates similar set-up times, tooling, and other machining requirements.

Once a family routing has been prepared and identified by the family number, a new part, when released and classified can "trigger" the issuance of the family routing to process planning personnel to prepare a specific routing for the new part. On an exception basis the "master" family routing can be marked up to customize the family routing for the new part and then entered into the routing data base. This method of preparing new routings is not only rapid, but will control standards and uniformity of manufacturing. When utilizing a computer and performing this task interactively, the system is even faster.

To further enhance family routings, usage of parts in such a family will determine whether it is economical to utilize group technology in manufacturing.

It is also practical to assume the addition of another type of classification in manufacturing called a *polycode* which will identify specific attributes of all parts regardless of the family from which they come. Such special attributes as milling, drilling, etc., mentioned previously can also be used when examining the potential of group technology.

PART STANDARDIZATION BY FAMILY

In previously determining preferred parts by family, such a definition will assist in preventing future unnecessary proliferation of designs. However, many current parts will be found to have variations in terms of tolerances, dimensions, views, instructions, and other characteristics. In order to reduce the cost of manufacturing current parts, it is appropriate to study families of parts with both variety and quantity to reduce the differences between them. This examination and change will reduce the cost of the current parts and also increase the quality of future parts as such standards are created. It should be remembered that few of the dimensions for a part are truly critical. The vast majority of dimensions are solely to define the part shape, and are subject to change in order that they may be identical to those of other parts, if practical. Thus, the ability to make such changes for uniformity and standardization is high. This is a most fertile field for cost improvement.

DEFINITION AND USE OF GROUP TECHNOLOGY

The following definition of *group technology* will assist in clarifying its use:

> *Group technology*   The analysis of similarly collected parts by shape, form, function, or operations, in a manner such that they may be manufactured in a similar manner using identical tooling or machine tools, either grouping the machine tools to combine a set of operations, or routing a family of parts over the same path to various machine tools.

The most popular and economical use of this procedure has been for parts with relatively low, "job shop" lot size. Implementing the procedure reduces set-up time for all the parts involved thus, in practice, mass producing small lots of parts.

The principle of group technology will eventually enhance material requirements planning also. It should be remembered that material requirements planning (MRP), develops requirements for parts through a planning horizon which extends out in time for one year or more. These requirements for parts are then individually examined and the requirements are combined on the basis of some economic lot size formula. These requirements create a "surplus" inventory which is, however, less expensive than the multiple set-ups which the combination of requirements eliminates.

Using the knowledge of a family of parts, it is now possible to combine requirements of a family of parts "vertically" in time, thus reducing not only set-up time but also the "surplus" inventory. This technique will create large savings in requirements planning.

CAD/CAM INTEGRATION

If all other phases of the implementation plan have been initiated and are working successfully, the integration of the classification system to CAD/CAM systems will be relatively simple. It should be emphasized that a proper data base for CAD/CAM is essential to its success. The classification system is the "glue" or the "paste" to make this occur efficiently.

The disciplines utilized in engineering and manufacturing will tend to become closer, and thus, the people and the input developed by different departments of a company may be combined. It is recommended that the organization be examined to make this closer interrelationship possible.

One form of organization may take the form of a "data services" department whose responsibility it is to prepare and input information into a data base for the following activities:

Engineering release data

Engineering change data

Classification and coding

Bill of material data

Producibility tips

Numerical control cutter center lines

Microfilm-microfiche system

Engineering standards

Manufacturing standards

Family routings

Discrete routings

In addition to this "text" type of input to a data base, drawings being prepared by either use of family drawing found through the classification system or automatically drawn using a design analysis program must be entered into the data base and stored. This large data base will be manageable if not only part data may be found by part number but also by family number.

Provision for additional data bases required for CAD/CAM must be made above and beyond those for business data processing. These include the following:

Part geometries

Model images

Design analysis data base

Family drawings

Family drawing programs

Producibility tips

Family routings

Family polycodes

NC family tool paths

Catalogs and cross references

NUMERICALLY CONTROLLED (NC) MACHINE TOOLS

No separate category was designated for numerically controlled machine tools previously. This does not, however, minimize their importance. A family of parts with similar form or shape imply identical tool paths with varying dimensions. Thus, a family tool path for the cutter center line will become important. Many parts will become conducive to design analysis followed by the automatic generation of such a tool path, including transfer to the proper postprocessor for the machine tool to be made. This activity, in effect will create the ability to make a part interactively with a machine tool.

Thus, provision must be made for family cutter center line data by family where appropriate.

FAMILY DRAWINGS FOR CAD/CAM

6

Many attempts have been made to reduce the time spent making drawings. For many years manufacturing companies have devised schemes to reduce line drawing time while continuing to convey adequate design data to manufacturing. With the advent of computer-aided design/computer-aided manufacturing (CAD/CAM), further impetus to reduced drafting time has become a significant goal. The principle of using family drawings holds great promise for achieving this increased productivity. It is the purpose of this chapter to explain the principles of family drawings, and suggest some guidelines for developing them.

The principle of family drawings is not new. Various methods for decreasing drafting time have been practiced for years. These include: (1) parts catalogs, (2) tabulated drawings, (3) "standard" parts, and (4) format drawings.

Until recently, a parts catalog has been the main source for critical dimensions and specification data for purchased parts. Many specific products appear in a catalog, and the product is dimensioned with "A," "B," or "C," dimensions. A list of the various product sizes or capacities also list the variable dimensions or capacities in tabular form. This method of design documentation is very popular and widespread.

Tabulated drawings, as a method of documentation, are, as we have already noted, very popular with engineers and drafters for designed parts. In this case a part with some specific shape or feature is drawn and all dimensions are again designated with A, B, and C dimensions as necessary. A table of parts with specific dimensions appears on the same drawing. Many problems have been found in the using this type of drawing, not the least of which is the continual lack of discipline in their use.

Format drawings are another method used to reduce drafting time. With these the major negative aspects of a tabulated drawing have been removed, in that, for a specific type of part, a complete drawing is prepared leaving only the dimensions of that part blank. In many cases certain standard dimensions are shown, such as finishes, fillets, etc. This has had a positive effect on the standardization of the drawings in a specific category. Unfortunately, format drawings have some disadvantages also. Often the drawing does not look at all like the finished part because, of necessity, the format drawing may be the "average" picture of a part which fits the drawing and does not show excessive lengths, widely divergent diameters, etc. Additionally,

127

some format drawings are used which, for example, may show a shaft with three diameters, and only two diameters are required for a new design for which the format drawing will be used. Thus, the picture is quite misleading. Finally, format drawings seem to be used most successfully when the product is fairly constant in design and shape such as a pump, electric motor, etc., and thus somewhat limited. For example, 300 format drawings can be used to cover probably 80% of the major parts of a pump.

Use of standard parts is another method which has been used to minimize drafting time. In this method, companies may prepare their own internal catalog of designed parts, tabulating the variable dimensions, and issuing the entire catalog to all parties who may need to know and identify them. Again, this creates all the problems of a tabulated drawing, some of those related to a format drawing, and also adds some more in that all parts in such a catalog are not necessarily current or even preferred. Such catalogs are often out-of-date, and do not reflect current and modern design. The indexes associated with such a catalog are often cumbersome, relying on the part name for retrieval, causing continuous interpretation.

With this background, what do we mean by part families or families of parts, and a family drawing?

As CAD/CAM becomes more and more popular, it will be necessary to re-evaluate how best to implement the electronic equipment involved to make drawings most economically. It is naïve to believe we can use an expensive cathode ray tube and allied computer equipment to make drawings more economically without some systematic effort to use the equipment. The CAD/CAM drafting system is the most expensive drafting board invented. Assuming the entire piece of equipment will be paid for simply by drawing lines faster seems a gross over simplification of the solution.

It is important to recognize that a cathode ray tube is, of itself, not the ultimate answer. We should set a goal to determine how many new lines we don't have to draw rather than how fast we draw new lines. If we set this as an objective, a family drawing makes sense.

We must assume a critical tool is available—a classification/design retrieval system. Regardless of the system used, parts with similar shapes will be grouped together such that one "family" drawing looks like all the parts in the group. There are exceptions to this as will be discussed later. Nevertheless, the basic premise will be to group similarly shaped parts, and, in some cases, those of similar function.

Figure 66 is the picture of a simple part. In this example, the family number is taken from a Brisch classification system having a code of 30011. Thus, it may be supposed that all round parts with a single diameter, with no throughgoing holes, and no external finishes, knurling, or threads will have this number.

Unfortunately, this is a family of several hundred similar parts. Thus, searching for a part identical to the one desired will be difficult and time consuming. Therefore, several other variables may be required to separate this large group of parts into reasonably smaller families. Length and diameter ranges as well as raw material specifications may be needed to so segregate.

As a result, conflict may occur simply because this necessary set of divisions for easy retrieval of a part may cause an unnecessary number of family drawings which look almost identical. In addition, those characteristics used to segregate parts into workable families may still omit other variables such as finish, and tolerances which can change the part sufficiently to create a new design. No scientific rule governs this

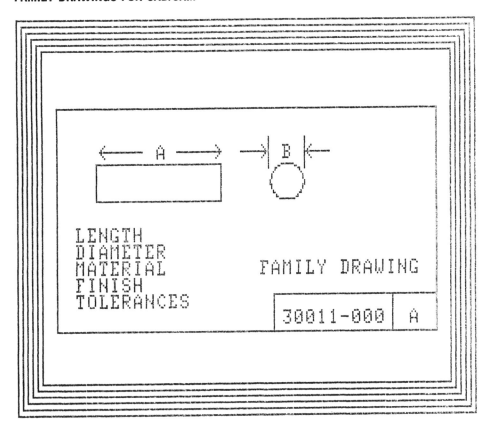

Figure 66 This simple shaped part might represent several families in a classification system.

conflict. However, if we look at how this type of drawing is related to a CAD/CAM drafting system, we may be able to establish some empirical rules.

We must suppose that stored in the data base of the CAD/CAM system reside all the necessary family drawings. Utilizing the classification retrieval system we are able to look at a catalog which contains all the parts in a family with the important "critical" dimensions listed in some sequence. Thus, the catalog section for 30011 should appear on the screen when called. If the families of parts have been segregated as mentioned previously by diameter and length, then a limited number of parts should appear on a list. Examining the catalog may determine that there is a part with the exact critical dimensions needed. However, it is not possible to use the part without first looking to see if it has additional characteristics unfavorable to its use which were not cataloged. If, after the specific part has been examined, an exact requirement is met, the search is over and the job is finished. If an exact fit is not found, it is necessary to design a new part.

At this time the family drawing will be recalled, and the critical A, B, and so on, dimensions replaced with the exact ones required. In addition, other data not cataloged must be added to completely describe the new part. Once the part is drawn, it is then released.

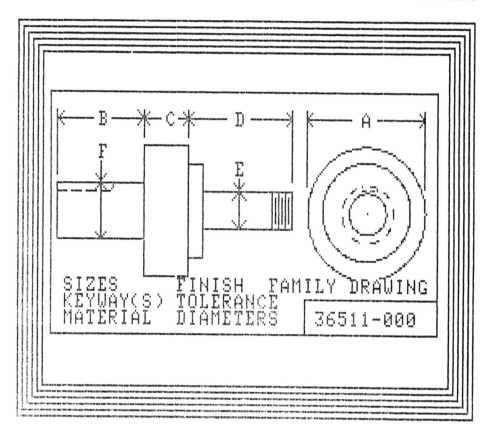

Figure 67 This fairly complex shaped part probably represents only one part family.

This procedure, particularly for more complex parts, is a significant timesaver. Lines are not drawn, or at least only a minimum need be drawn. However, it is here that empirical rules will help. If the "add ons" create sufficient additional drafting time so as to be "excessive," then an additional family drawing will reduce this add-on time. If too many family drawings are made to reduce add-on time, too many family drawings must be stored and retrieved for efficient management of the system. It would appear that "simple" parts as in this example may have many families with one family drawing.

Figure 67 is an example of a more complex part. As in the case for all family drawings, all of the specific features to be found in any one family are not cataloged. The characteristics are also not specifically designated on the family drawing. This will be true in any and all families of parts. However, in this example, the complexity of the part is far greater, and the drawing time is large compared to the minor variations which may be found on the parts. Thus, it would seem that one family drawing will be required for each family as the family of drawings becomes more complex. The dividing line between parts such as those shown in Figures 66 and 67 will require some analysis.

As the parts become more complex, it is also evident that a large amount of time can be saved by using family drawings and replacing generic alpha dimensions with actual dimensions. An increasing number of lines do not have to be drawn. On very complex drawings much time is spent simply determining how to layout the rows of dimensions to adequately define the part. Once a family drawing has been prepared this tedious task is finished hopefully forever.

Not mentioned previously, but a positive feature of most CAD/CAM systems is the ability to make a new drawing to scale using a customized computer program. Such programs would be prohibitively expensive if required for each new part required. When written for a family of parts, however, the economics of such a program become reasonable, adding a really exciting attribute to new drawings – a drawing once again which has been made to scale in all dimensions.

Figure 68 shows an example of another type of problem encountered in family drawings. Many parts cannot be defined and classified by form or feature, but only by function. In this illustration a family of "brackets" are shown. None of them look alike, and no single family drawing could hope to be used to draw them. If an attempt was made to create a single drawing for them, it would soon be apparent that so many alterations would need to be made to suit the next bracket, that it would be more economical to draw a new one. Thus, it is inevitable that some families of parts may have many family drawings to describe the parts in one family. This, too, will take some study to determine where the boundary for such families begins. Under these

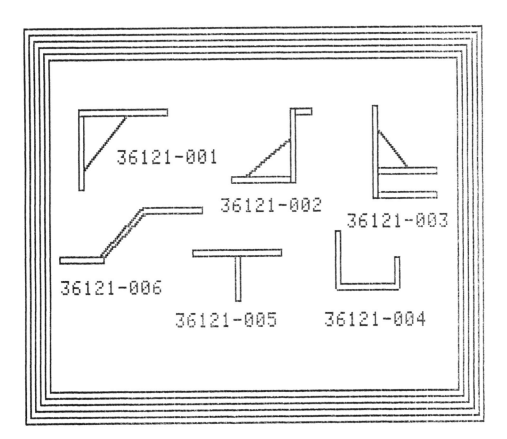

Figure 68 These parts in one family may each require a separate representation.

conditions, it will also be somewhat difficult to prepare a logical catalog. Thus, it may well be that the only way to search for a part to see whether it exists will be to look at each actual drawing in the family. This will add a requirement to the CAD/CAM system that it be able to view, in miniature, several parts in the same family on the screen at the same time. While this feature may be useful for all families, it will be almost mandatory for parts described by function as shown here.

Figure 69 is the last example of part family types, or assemblies. The significant variety of such families may, once again, require that each finished design be considered a family drawing. However, such assemblies as electric motors, pumps, gear trains, compressors, etc. may prove conducive to create one family drawing.

Figure 70 is an attempt to help lead those who study this problem onto a path where empirical rules can be prepared and be beneficial. Briefly, the graph points out that the fewer variations allowed to be placed on the family drawing as add ons, the more family drawings are required. At the same time, when more family drawings are made, more variations can be placed on the family drawing. Somewhere an optimum exists which minimizes the total population of family drawings as opposed to the amount of time necessary to add the variations not covered.

The family drawing takes the best attributes of all the old ways for reducing drafting, and leaves out the bad practices:

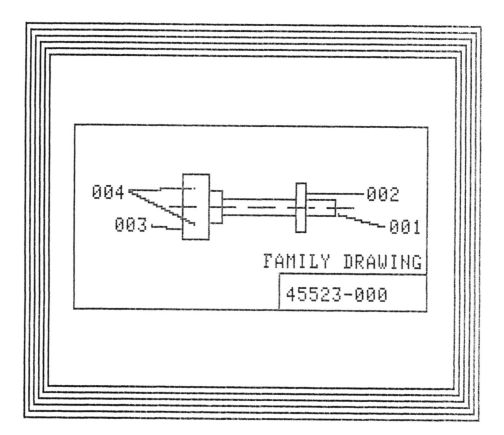

Figure 69 Families of assemblies do not lend themselves as well to general drawings.

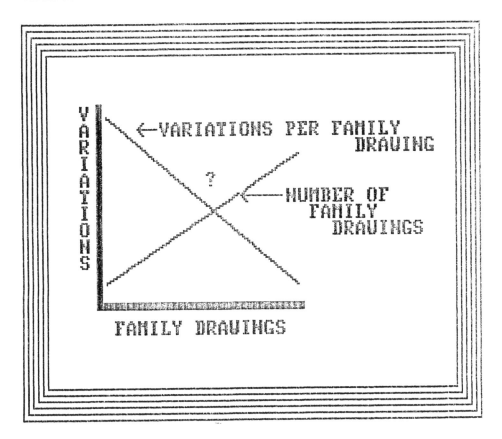

Figure 70 When determining the number of general drawings required, as the number of variations allowed on the drawing decrease, the number of family drawings will increase.

1. Each part has a separate document eliminating the confusion of the tabulated drawing.
2. The classification system brings together parts so closely similar that simplification and standardization can be achieved.
3. The classification system helps to avoid creating duplicate parts, or unnecessarily close similars.
4. The standard parts catalog with noncurrent data is eliminated because the classification system is always current when using CAD/CAM. It may also cover more or perhaps all rather than limited categories of parts.
5. The format drawing principles are completely incorporated, but can now also be drawn to scale with CAD/CAM.

Another condition should be studied in the use of family drawings concerning the type of drawing required for each. Three different drawing formats are possible in the use of CAD/CAM. Two-dimensional (2-d), flat drawings are required for such things as wiring diagrams, electronic circuitry, and footing layouts. Three-dimensional (3-d) or wire-frame drawings are required for the majority of part families so that sufficient views of the part may be created. However, some families will require a geometric model where surfaces become important. This more complex requirement is due to the need for design analysis such as finite element analysis, gear design, and others. Also, those parts which will eventually be manufactured on a numerically controlled machine tool need surface description as a geometric model. To make full use of family drawings, each family should be examined for these requirements and incorporated into the required drafting standards for the family.

It is interesting to note that this integration of requirements further enhances the economics of making drawings and allows the same data to be used by all departments requiring it.

Thus, three types of family drawings are possible:

One drawing may represent parts in several families

One drawing may represent parts in one family

Several drawings may represent parts in one family

GENERAL CONCEPTS OF A PRODUCT IDENTITY SYSTEM

7

THE PARADOX OF INTEGRATED SYSTEMS

As we begin discussing product identification systems and all the systems surrounding them, we must be aware of some of the paradoxes they cause. Modern systems imply data processing. Data processing implies integration of systems and departments. It is these two factors which cause the paradox.

Figure 71 shows a typical company organization chart. In each segment such as engineering, manufacturing, or accounting, systems exist, even though they may be only the utilization of paper and pencil. These systems, however, tend to work "vertically." That is, most systems operate internally within a segment of the company simply because it is managed as an entity. The white spaces between these vertical rows of blocks are really walls. One dares not venture to the next departmental area. It is somewhat similar to the feudal baronies of the Dark Ages. Each serf was given a plot of ground on which he could plant crops in return for his allegiance to the lord of the manor. Woe be he who would travel to the next barony and communicate with "the other crowd."

As we proceed with this chapter, our thinking will be directed toward an integrated system. The product identification set of activities directly affect every other department of a company in the form of bills of material, product structures, item master data, configuration management, and designs represented by drawings. Figure 72 is a simplified organization chart showing how business systems are tied together. In essence, the systems tend to flow "horizontally" through a company with no regard for the artificial (even if necessary), organization. These barriers must be softened or eliminated in order for an effective product identification system to operate effectively. Another problem has been how manufacturing companies consider a product identification system and its included bill of material system. Often, manufacturing personnel, or data processing personnel, in desperation, decide to take the manual output of an engineering department and "mechanize it." To their horror, they run into all sorts of problems, omissions, communication deficiencies, and lack of back-up, or poor documentation.

Figure 73 shows a manual bill of material being planned for mechanization by

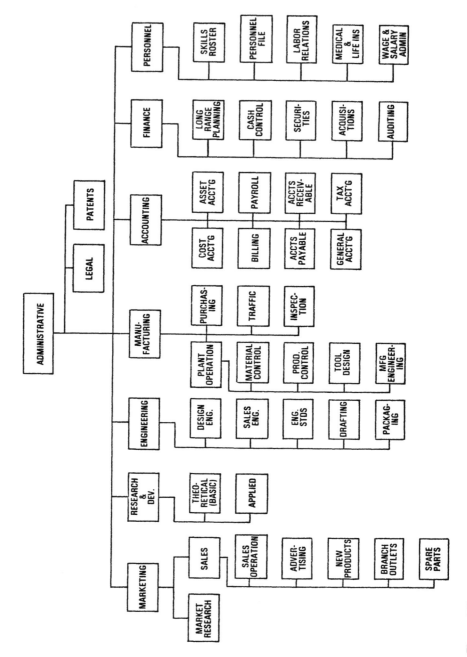

Figure 71 A highly stylized company organization chart is shown here to illustrate the vertical relationships within.

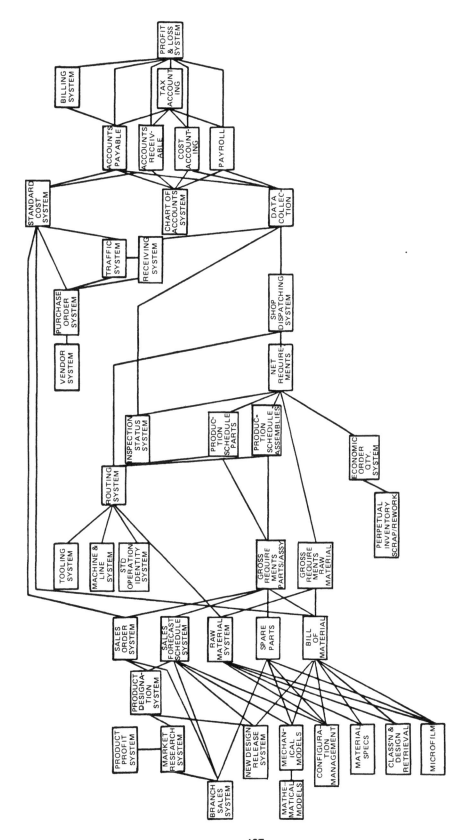

Figure 72 Systems utilized by a firm are "horizontal" in that they inevitably cross organizational lines.

137

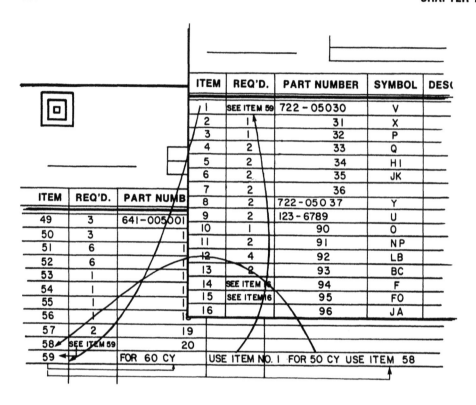

Figure 73   This "involuted" bill of material has so many instructions in the notes it is possible to go in circles. The engineer probably knew what he meant. The disciplines of mechanization of a bill of material system will clean up these problems. (They had better, or chaos will result.)

data processing personnel of one company. Let us look at some problems data processing might encounter during the process.

The bill of material example has two pages. The first three columns on each page contain item number, quantity required, and part number. Let us assume we are in manufacturing preparing to assemble the unit. As we look down these three columns, the very first item shows a reference which says, "see item 59." In all innocence, we adjust our eyes to item 59 on the second page. At this point we raise our eyebrows slightly, for we are now told, "for 60 cycle, use item 1, for 50 cycle use item 58." We sigh with relief as we remember the unit to be manufactured is, in fact, 50 cycles, and thus we won't have to make continuous circles between item 1 and item 59. Thus, we cast our eyes hopefully onto item 58. To our chagrin, we find our next instruction is, "see item 59."

Finally, we must face the fact that no one rigid system will work effectively all the time. Figure 74 is a facsimile of an implementation schedule. Please note two lines of activity in particular. One line states, "bill of material, standard production." Another line states, "bill of material, nonstandard production." These two common sense activities recognize that, in times of emergency, even the most dedicated person will bypass a system. If provision for so doing is not recognized and documented, the sys-

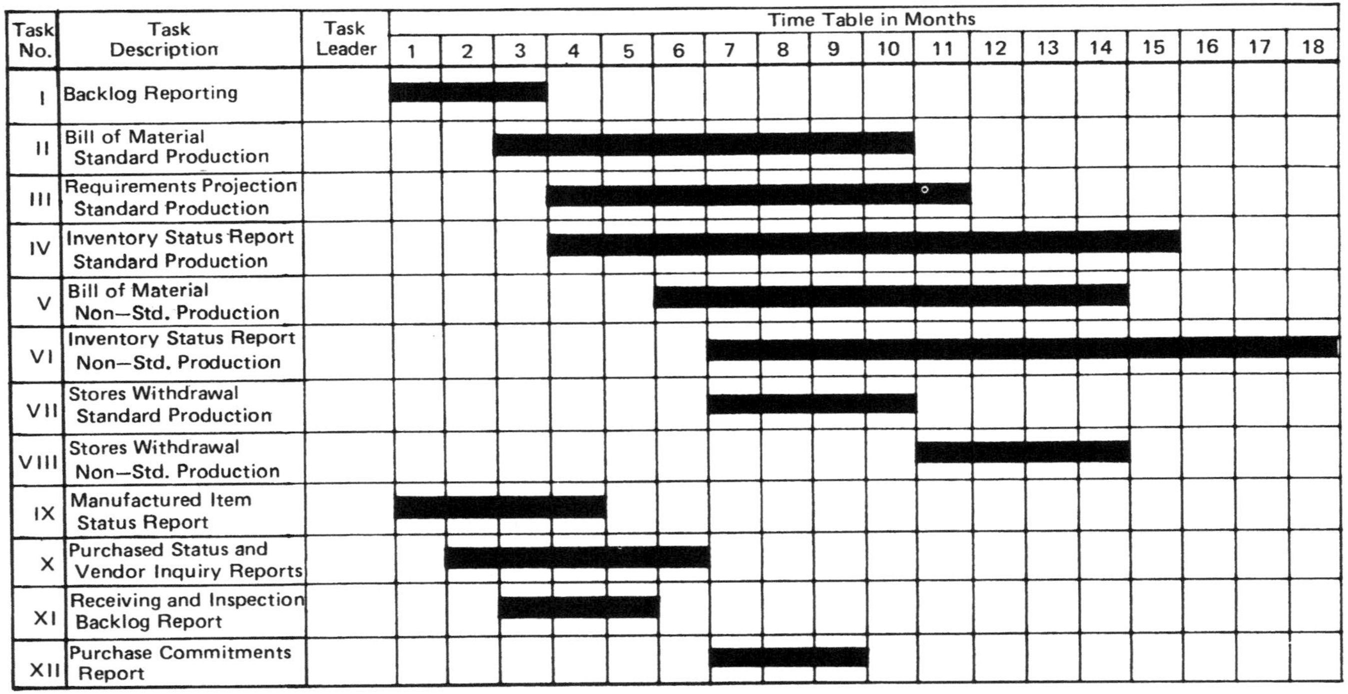

Figure 74 Not only is a plan and schedule required to implement a system successfully, alternate solutions for variations to the problem are necessary. (Reprinted from *Management Controls*, copyright 1970, Peat, Marwick, Mitchell and Company, New York, New York.)

139

tem will be full of errors and omissions. Thus, even as we begin to discuss some of the rules which affect all business sytems, let us be aware that common sense may alter some of the rules part of the time. We must know where and when this might occur and plan accordingly.

PRINCIPLES

Two principles are involved in a successful product identification system, namely part and assembly "identity," and product and assembly "structuring." Utilizing the rules associated with these principles correctly, with necessary variations to conform to the unique problems or procedures of a specific manufacturing company, will provide the necessary tools to control engineering design documentation. Integrity of the system will consequently add to, or significantly improve production control, material requirements planning, inventory control, purchasing, and accounting systems.

RULES

Identity

Each part and assembly should be designated by a unique identity called a part number. The part number should preferably be 8-digits or less in length, nonsignificant, containing no suffixes or prefixes, and may not include the revision-level number or letter as part of the part number. In addition, the following statements or rules are to be followed:

1. The drawing number and part number are the same.
2. All drawings will be assigned and identified by a part number.
3. Some parts and assemblies, identified by part numbers, may not require a drawing.
4. Rough forgings, castings, and raw materials have part numbers other than that of the finished part for which they will be used.
5. A part number for a part or assembly will not be changed for any artificial or external reason such as when a part is changed from "nonstock" to "stock," or "make" to "buy," etc.

Structuring

Structuring encompasses the method for picturing how parts and assemblies are put together to make a complete product. A product structure is defined as a graphic or pictorial "schema" showing the hierarchical relationship and combinations of parts, assemblies, and subassemblies in a product. These relationships portray the manufacturing assemblies and "levels" of each part and assembly. We will discuss this relationship in more detail in chapter 8.

Figure 75 however pictures an important and significant equality in a product structure. Each assembly drawing should be based on a predefined "general" product structure. This general product structure is converted to a detached assembly parts

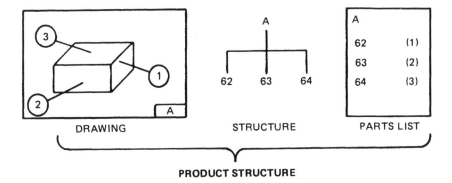

Figure 75 Equality between an assembly drawing, a specific product structure, and the corresponding parts list is mandatory.

list. Thus, we can define an equality between an assembly drawing, a specific product structure, and a parts list.

Two types of product structures are therefore utilized:

The *general product structure* is developed for a product line to create a typical part–assembly relationship. A general product structure identifies parts and assemblies generically by name, not part numbers, as it represents the entire product line part-assembly relationship.

A *specific structure* is developed for each unique design, or customer order. The specific components documented for each assembly and subassembly are determined by locating their position on the appropriate general product structure, and recording the appropriate "parent," "component" relationship. This documentation will consist of a detached parts list for each assembly, showing the parts and/or subassemblies required.

Be aware of the rule, "all drawings must have a part number, all part numbers may not require a drawing." Some instances, such as "kits," meet the latter part of the rule.

PIECE PART AND ASSEMBLY DRAWINGS

As a result of these concepts, specific documentation and additional rules are required to perpetuate and control the system. Important sets of documents are part and assembly drawings.

PIECE PART DRAWINGS

Piece part drawings should be made for every part. Each part will have its own individual drawing. A piece part drawing will show only one item made from one material, and contain only that information required to make the part. This information will include material (commercial or company material specifications), and any special tolerance or instructions to make the part. The instructions should be of a "what" nature, such as copperplating, hardness specification, etc., and now "how" instructions which might describe or limit the manufacturing process.

An engineering standard should be written describing in detail the drawing content requirements.

The rules concerning piece part drawings sound innocent. We have discussed previously the need for piece part drawings, and now must define more clearly drawing contents.

ASSEMBLY DRAWINGS

Any combination of two or more parts which create a new entity is called an assembly. This assembly will have a part number which is also the number of the assembly drawing upon which the design appears. The assembly drawing will include only that information required to assemble the parts, including dimensions, plus any special tolerances or instructions for assembly. The components of the assembly will not be listed on the face of the assembly drawing, but will be maintained on a detached parts list.

Figure 76 is an assembly. A pump impeller has been used as an example. Find numbers appear both on the drawing and the parts list so cross reference from one to the other may be maintained.

The parts for an assembly are identified on the assembly drawing by use of find numbers. These numbers are needed and used to link the parts on the assembly drawing to the assembly parts list or the other way around. No parts list appears on the drawing, but, rather appears on a separate document called the *detached parts list*. Hence the need for find numbers.

Find numbers are either numeric; assigned sequentially as 001, 002, etc., or alphabetic as A, B, C, etc.

Figure 77 shows an example of a beautiful old drawing which, in a modern system, would have to be altered or replaced. Many companies still use this kind of drawing with the attached parts list. Those companies who change will X out the parts list, and transfer it to the detached parts list.

Most companies who change systems recognize the need to inform all designers and drafters of the new methods required to make a drawing. Ingersoll Rand, for example, developed a miniature booklet for reference. A segment is shown in Figure 78. The book is small enough for a shirt pocket and turned out to be quite popular with those who must use it.

PART NUMBER LOG

As each new design is assigned a part number, the part number and associated data should be entered into a *log*. Since much significant data may have been incorporated into a former part number system, the data, detached or eliminated for the new part number system should also appear in the log.

It should be emphasized that, as in this example, a mechanized system may change the method and location of retaining data. For example, the part number log might well be a part of the mechanized system. What we will later describe as the item master may be the repository of all data associated with the part number. Circumstances will dictate the size, amount, and need for manual systems.

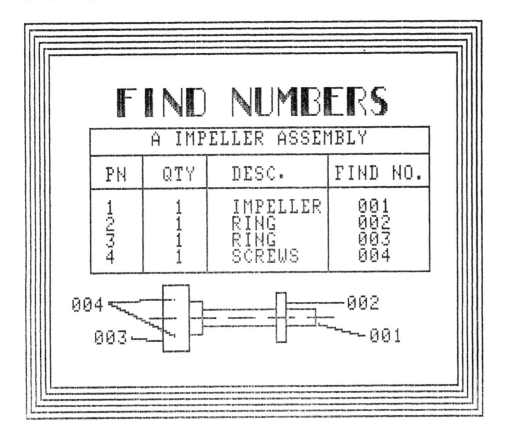

Figure 76 A detached parts list is utilized to correspond to a specific assembly. Find numbers are used to correlate the two documents.

It is quite feasible and practical to mechanize the log so that it may be sorted by any design criteria or significant piece of data required, thereby making it possible to retrieve similar parts. The release procedure, described later, may replace this log, particularly when a mechanized bill of material is implemented.

CONFIGURATION MANAGEMENT

A system incorporating the use of part numbers, assemblies, and bills of materials requires procedural formalities of rules generally called configuration management and control. Engineering standards and procedures are required to provide guidelines such that documentation integrity may be maintained.

Configuration management is defined as:

Configuration management The systematic approach to identifying, controlling, and accounting for the status of the parts and assemblies required in a contract and/or design from the point of its initial definition throughout its entire intended life.

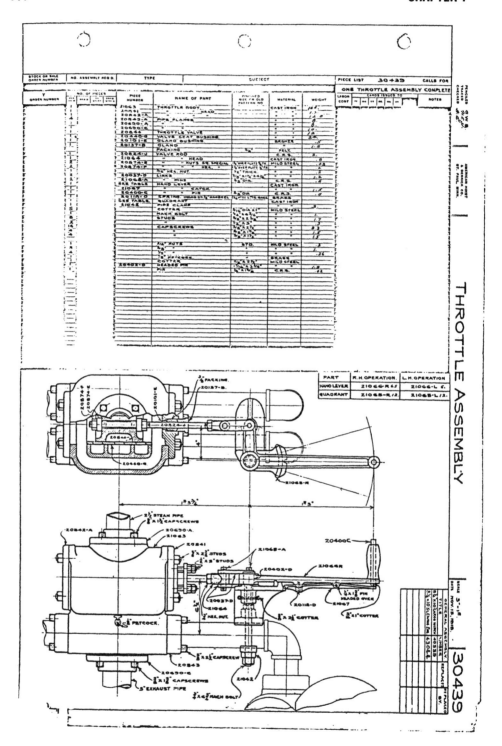

Figure 77 This meticulously drawn assembly has insufficient data to be mechanized. (Courtesy of American Hoist & Derrick, St. Paul, Minn.)

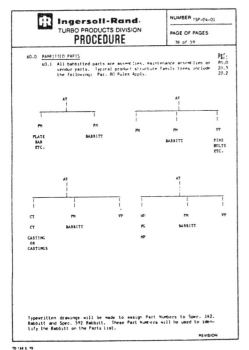

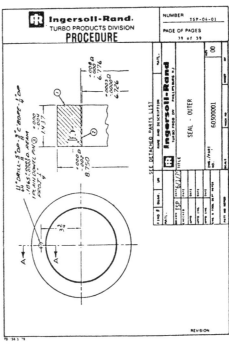

Figure 78　New standards to make it possible to mechanize an engineering system require the instructions to be easily available. These sample pages come from a shirt-pocket-sized procedure manual. (Courtesy of Ingersoll-Rand, Turbo Products Division, Phillipsburg, NJ.)

To manage configuration requires clear and concise rules on drawing content, part number issue, and their record of revision-level changes. Knowledge of when a new part number must be issued as opposed to a revision level change of an old part number is vital.

Figure 79 shows the basic ingredients of configuration management, namely a product structure, or family tree, released drawings, specifications, bills of material, manuals, spare parts, and the definition of a standard product with options. Behind these basic requirements are a series of necessary policies, standards, and procedures. Perhaps, if we illustrate the flow of data, the need for these written rules and their sequence may be discerned.

Flow of data and responsibility for issuing, maintaining, and altering such data is necessary to create confidence in the system and eliminate the need for generating parallel and redundant systems to check the main system. Policies, standards, and procedures are thus required.

A system is dynamic and also has exceptions. Provision must also be made to alter it in an orderly and flexible manner.

Configuration management can be tedious. Perhaps this is why so many companies have mediocre document control. We shall try to use illustrations to clarify the concept, and also to attempt to prove that the principles are straightforward.

A new product activity may be divided into four sections, a functional definition,

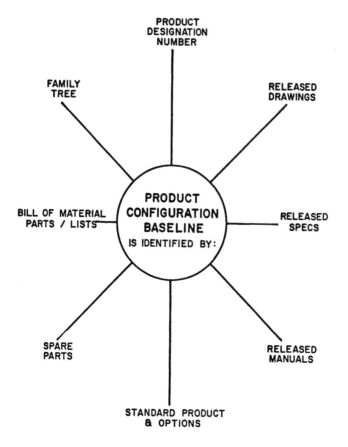

Figure 79 Configuration management is a documentation control. This control is based on the items identified in this figure.)

the *hardware* definition, the manufacturing definition, and the field definition. As a new product passes from the design stage, through a prototype manufacture, to final production, control of the configuration of the product is highly important. In Figure 80 three phases of product development are shown. Under each appear various names of configuration control requirements. Three blocks appear under each title representing policy, standard, or procedure. Notice that the title "configuration management" appears under the functional definition, product or system plan. Both a policy and a standard are required. We find 8 specific requirements under the functional definition, 1 under hardware definition, 12 under manufacturing definition, and 1 under field definition.

The requirements pictured have been consolidated into a chart as shown in Figure 81. Also combined on this chart, are all the departments of a company influenced by configuration management and control. They are as follows:

Marketing

Product line management

Engineering

Publications

Software (if computers)

Manufacturing engineering

Production and inventory control

Manufacturing

Purchasing

Quality assurance

Customer engineering

Accounting

Data processing

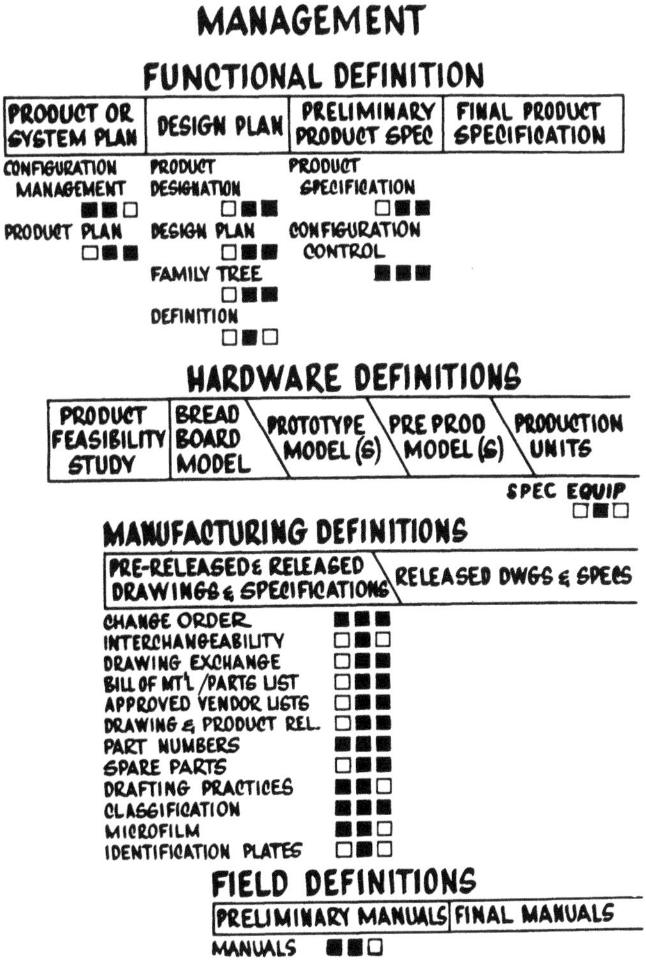

Figure 80 Configuration management is further defined in three stages of design development, the functional, hardware, and manufacturing definitions.

		REQUIRED			DEPARTMENTS AFFECTED												
		POLICY	STD	PROCEDURE	MKTG.	PROD.LINE MGT	ENGINEERG	PUB	SOFTWARE	MFG. ENG.	PROD.INV.CONT.	MFG.	PURCH.	Q.A.	CUST. ENG.	ACCTG	DATA PROCESS
1	DRAFTING STD.	N	Y	N	X	X	X	X	X	X	X	X	X	X	X	X	
2	FAMILY TREE STD.	N	Y	Y			X	X		X		X				X	X	X
3	PRODUCT PLAN STD.	Y	Y	Y	X	X	X		X		X			X		X	X	X
4	PRODUCT DESIGN STD.	Y	Y	N	X	X	X	X	X	X			X		X	X	X
5	PRODUCT DESIGNAT. STD.	Y	Y	N	X	X	X	X	X						X	X	
6	PERT STD.	Y	Y	Y	X	X	X	X	X	X	X	X	X	X	X	X	X
7	CONFIGUR. MGMT. STD.	Y	Y	Y	X	X	X	X	X	X	X	X	X	X	X	X	X
8	DRWG. & PROD. RELEASE STD.	N	Y	Y			X	X	X	X	X	X	X	X		X	X
9	MICROFILM STD.	Y	Y	Y			X	X	X								X
10	DRWG. EXCHANGE STD.	Y	Y	Y			X	X	X	X	X		X		X	X	X
11	BOM/PARTS LISTS STD.	Y	Y	Y			X	X	X	X	X	X	X	X	X	X	X
12	SPARE PARTS STD.	Y	Y	Y	X	X	X	X		X	X		X	X	X	X	X
13	QSE STD.	Y	Y	Y	X	X	X	X	X	X	X	X	X	X	X	X	X
14	SPECIAL OPTIONS STD.	Y	Y	Y	X	X	X	X	X	X	X	X		X	X	X	X
15	PROD. SUPT. MANUAL STD.	N	Y	Y	X	X	X	X	X			X	X	X			
16	INTERCH'G'ABILITY STD.	Y	Y	Y	X	X	X	X	X	X	X	X	X	X	X	X	X
17	PART & DOCUMENT STD.	Y	Y	Y	X	X	X	X	X	X	X	X	X	X	X	X	X
18	CLASSIFICATION STD.	Y	Y	Y			X	X		X	X	X	X	X	X	X	X
19	APPR. VENDOR STD.	Y	Y	Y			X	X			X		X	X	X		X
20	ENG. DES. - MG. PROC. STD.	N	Y	Y			X			X		X	X	X			
21	CONFIG. CONTROL STD.	Y	Y	Y	X	X	X	X	X	X	X	X	X	X	X		X
22	FIELD CHANGE ORDER STD.	Y	Y	Y	X	X	X	X	X		X		X	X	X	X	X
24	ENGINEERG. CO. STD.	Y	Y	Y	X	X	X	X	X	X	X	X	X	X	X	X	X
	TOTALS	18	23	20	15	15	23	20	18	18	16	30	20	17	20	19	18

Figure 81 To control original documentation of designs as well as changes, a series of policies, standards, and procedures are recommended. The "Y" in the appropriate columns shows what is recommended. The "X" in the appropriate columns shows whether that particular document affects departments in the company.

Often a company with system problems attacks the "symptoms" instead of the cause. Excessive inventory, "hot lists," extensive rework, late deliveries; each may seem to be the problem. More often, configuration management is not working. Figure 82 is an attempt to show the relationship between portions of configuration management to such problems. The following problems can be related to the corresponding standards and procedures:

Duplication of designs
    Classification
    Drawing exchange
    Microfilm
Change order problems
    Engineering change order
    Bill of material
    Release procedure

Interchangeability

Configuration control

Part and document numbers

Equipment contents

Drawing content

Product structure

Product plan

Design plan

Product numbering system

Special equipment

Approved vendor definitions

Product Specifications

Servicing

Spare parts definition

Parts manuals

Field change orders

PRESENT PROBLEM

SOLUTION

DUPLICATION
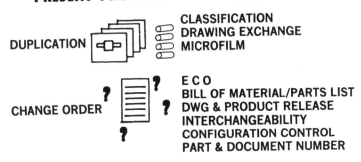

CLASSIFICATION
DRAWING EXCHANGE
MICROFILM

CHANGE ORDER

E C O
BILL OF MATERIAL/PARTS LIST
DWG & PRODUCT RELEASE
INTERCHANGEABILITY
CONFIGURATION CONTROL
PART & DOCUMENT NUMBER

EQUIPMENT

DRAFTING
FAMILY TREE
PRODUCT PLAN
DESIGN PLAN
PRODUCT DESIGNATION
SPECIAL EQUIPMENT
APPROVED VENDOR
PRODUCT SPEC.

SERVICING

SPARE PARTS
MANUALS
FIELD CHANGE ORDER

Figure 82 Several engineering documentation problems can be identified as lack of good configuration management.

If, as portrayed in Figure 83 product costs are increasing, disproportionate delays are occurring in making engineering changes, the product leaving the assembly line contains unknown parts or assemblies, the product is not built as engineering documents it, and there is consistent downtime in manufacturing waiting for resolution to these problems, then configuration management had better not remain a boring or dull subject in your operation. From a documentation flow viewpoint, this can be the most important and effective subject for study by any company today, requiring immediate managerial attention!

REVISION LEVEL

Figure 84 shows a drawing with a proper revision level applied.

Design changes require either a new part number to be issued with a corresponding new drawing, if applicable, or a revision change to the old drawing.

All new designs, unless *prereleased,* enter the system with a revision level of 01, (or A). Design changes which do not affect the interchangeability of the old design and the new design are indicated by changing the one- or two-digit revision level sequentially from 00 to 01, etc.

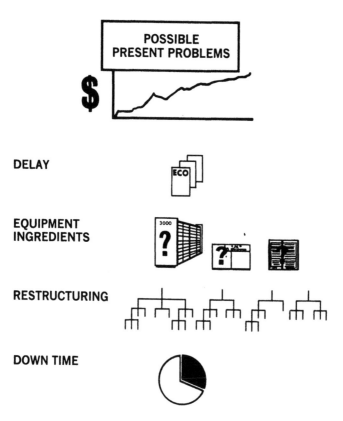

Figure 83 Additional problems in engineering documentation due to lack of good configuration management are illustrated.

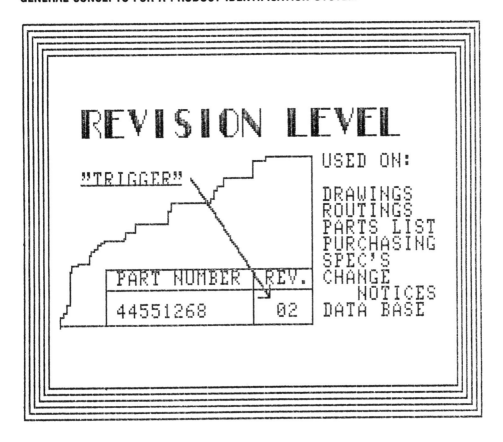

Figure 84   The revision level of a drawing is used as a "trigger" to inform people of an engineering change. It is not a portion of the part number.

It must be emphasized that the revision level is not included in the part number. Consequently, a part or assembly stocked in inventory by part number may possibly be found in the storage bin with more than one revision level in effect.

The revision level is used as a "signal," or tracking device, to record and catalog changes which do not affect interchangeability, and which, after implementation, are not needed to identify the part.

The revision level is used and/or tracked on the following documents:

Engineering drawings

Specifications

Purchased part drawings

Purchase orders

Routings

Engineering change notice

Item master file

Assembly parts list (for the assembly part number only)

INTERCHANGEABILITY

Interchangeability rules have been developed to assist a designer or drafter determine when a new part number must be issued for a design as opposed to changing a revision level of an old design. Figure 85 is a decision table which will prove helpful in making this decision.

The more formal definitions or rules are as follows:

Interchangeable   When two or more parts possess such fundamental and physical characteristics as to be equivalent in performance and reliability and capable of being exchanged one for the other without alteration of the parts themselves or any adjoining parts except for normal adjustments, and do not require selection for fit or performance, the parts are *interchangeable*. A change to a previously designed part which still allows it to be used according to this rule shall have a revision change only

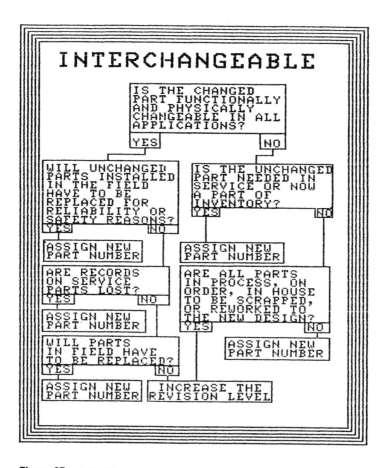

Figure 85   A decision table assists the designer to determine whether a part number or revision-level change is required.

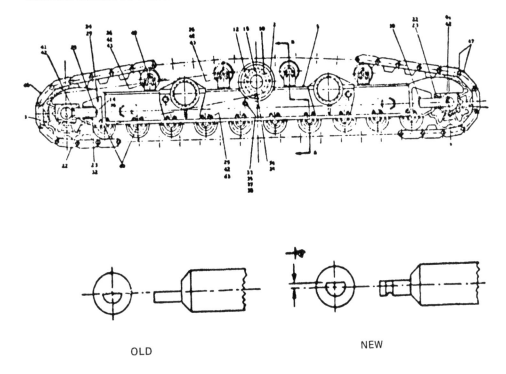

OLD

NEW

Figure 86 When a part is to be changed and the new part will not replace the old in function, fit, or durability, a part number change must occur. This example of a design change where the two designs are not interchangeable illustrate the often complex ingredients of the decision to change a part number or revision level. (Courtesy of American Hoist and Derrick, St. Paul, Minn.)

Noninterchangeable When a part is to be changed and the changed part does not physically fit or will not function as a replacement for the unchanged part in all applications of its previous revisions, and all unchanged parts will not physically fit or function as a replacement for the proposed change, the parts are considered *noninterchangeable,* and a new part number and corresponding drawing is issued for the new design.

Figure 86 is a picture of a major assembly in which the shaft for the idler wheels has been replaced. The new shaft has a ¼ inch offset, which is different from the old shaft. Should this part have a new part number?

It is true that either shaft will operate correctly in the assembly—as long as all the shafts are identical—. However, it would be impossible to determine which shaft is in the field on a particular unit without a part number change. Consequently, a new part number is required.

It is possible to visualize many examples which seem complex. However, when faced with a real example, common sense solves the problem rather easily. It is important that each engineer, designer, and drafter follow the rules of interchangeability, or else the recipients of design documentation will have little confidence in such data. This set of rules cannot be emphasized too greatly. They are imperative!

ENGINEERING CHANGES

Engineering changes must be documented and communicated throughout the functional areas of engineering, manufacturing, service, and accounting. Figure 87 pictures all of the various activities affected by an engineering change. Certainly it is not just the drawing or the bill of material, but a host of documents and production planning functions which are influenced. Strangely enough, many companies still have no formal change notice procedure.

There are many methods and different forms used to convey engineering changes. Figure 88 pictures a typical, but stylized, change notice form. The system must convey the type change, the severity, the date or serial number on which the change will be made or "effectivity," and the disposition of old parts or assemblies. If the severity criteria requires a cost estimation, this data should also appear on the change.

It may have become a status symbol to receive a copy of an engineering change notice. Certainly approval of one usually requires a substantial number of signatures. We must find a way to minimize the turnaround time for engineering changes, for they can stop production until they are resolved. Completion of most engineering change approvals should be made within 24 hours.

Most companies do not discriminate between a severe and a simple change. The following set of definitions, used with a change notice might serve to help in the routing of the change and also in determining authorization procedures and speed of approval. At first glance, these may seem too elaborate. On the other hand, each subject listed is part of the information needed to make a sensible decision about a change.

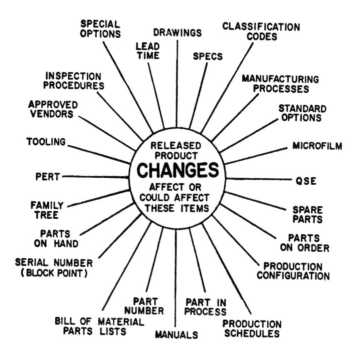

Figure 87 An engineering change initiates a chain of additional activities. Without a coordinated plan, confusion and lack of communication can cause serious problems.

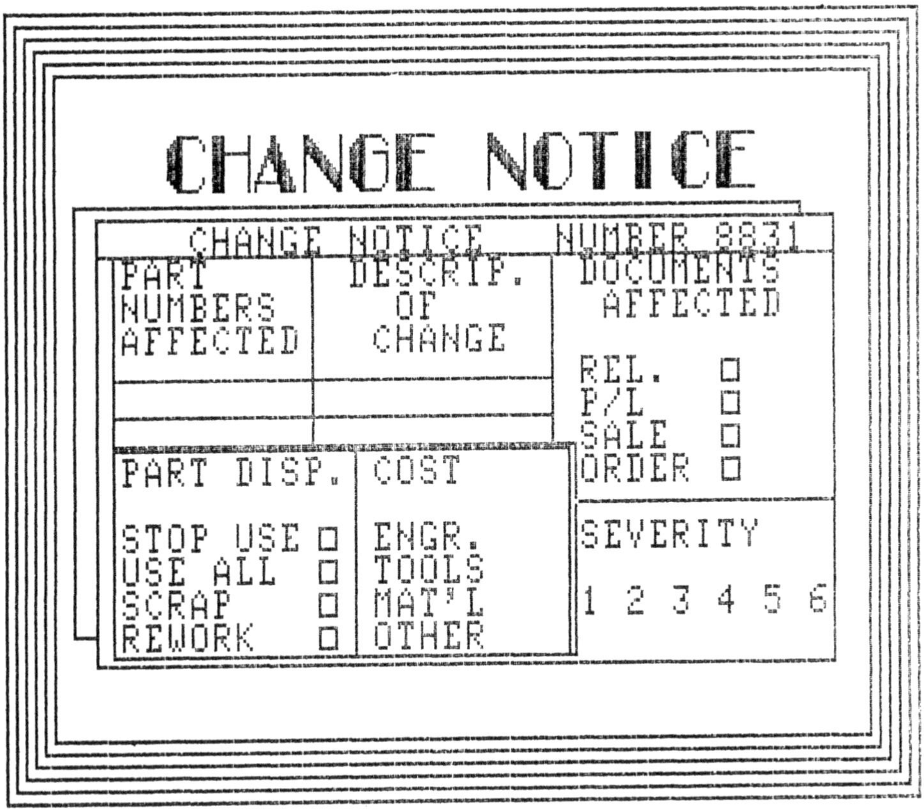

Figure 88   The sample engineering change notice shown here covers the major ingredients necessary to communicate a design change to all departments.

Thus, if the data listed here can be recorded and transmitted efficiently, information and facts for decision will be on hand.

ENGINEERING CHANGE NOTICE CLASSIFICATIONS

Type
  Documentation change, affects price
  Complex change, coordinates several changes
  Part or assembly change, changes next level(s)
  Simple part or assembly change
Status (of part being changed)
  Prereleased
  Prototype
  Released

Severity

    Customer change affects parts in inventory, on order,

    In field affects parts in inventory

    On order affects parts in inventory record change only

Urgency

    Emergency

    Routine

    No change board activity

Priority

    Customer required

    Line shutdown

    Rejected material

    Cost reduction ($>$\$500/year)

    Significant design improvement

    Minor documentation correction

Reason

    New industry demand

    Customer service

    Product refinement

    Customer request

    Manufacturing adjustment

    First issue

    Engineering standards

    In addition, documentation changes must conform to sound configuration control rules and rules of interchangeability. Figure 89 shows a summary of the type changes possible. Each change is documented using the rules of interchangeability.

    Let us look more closely at each of the three possibilities. You can see a certain logical progression in the changes as they go from simple to complex.

    The simplest change is shown in Figure 90. Part 1, part 2, and part 3 are assembled into part number A. In turn, assembly A is a portion of higher-level assembly X. The subscripts 0 are intended to denote the revision level of each part number, and are all 0 as the example is first documented.

    A change is made to part 1, such as change of finish, addition of a tolerance, or some other alteration which does not affect its interchangeability. As shown in the figure, part number 1, revision 0 is altered to part number 1, revision 1.

    The documents affected will be the change notice, initiated for part one, and the drawing for part 1. The date of change is not necessarily important, since both revision 0 and revision 1 are interchangeable. Of course, if a cleaning or deburring operation was added, then the *time* of change will become important. "Next order" is usually sufficient timing.

    Figure 91 shows a more complex change. In this example a change has been made

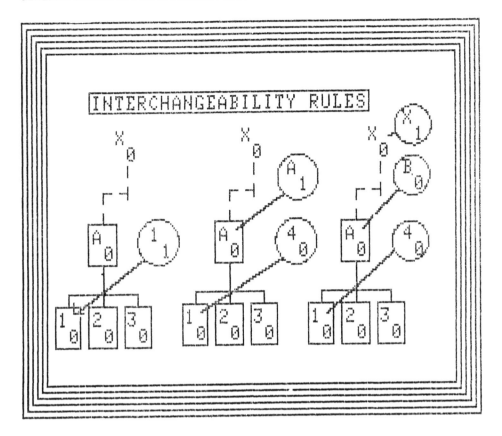

Figure 89   The affect of a design change varies from the very simple to the comparatively complex. The steps, however, are logical and progressive.

again to part 1, but the change is noninterchangeable, causing a part number change. Thus, the release of a new part—part 4.

Notice, however, that the change of part 1 to part 4 has required investigation of the interchangeability at the next higher level, part A. By implication, interchangeability has been maintained for part A, but the revision level must change to record the fact that some change has occurred to the assembly.

The documents affected are the change notice, initiated for part A, the drawing itself, changing the revision level on the drawing, the detached parts list, and a new part release notice for part 4.

It should be emphasized here that in the first example, shown in Figure 90, no revision level of any kind had to be changed on a parts list, because the revision levels of the components do not appear there. Even when an assembly is listed as a subassembly on a higher level, the revision level is not posted. Only the "parent" part number of an assembly has a revision level posted.

Figure 92 is the most complex of all changes, one which could affect all levels of a bill of material.

Once more a change has been made to part 1, replacing it with part 4. In this case, however, the change creates a noninterchangeable assembly. The result is a part number change from part A to part B. The implied rules we have been describing continue to take affect. Since part number A has changed to part number B, we must look to the next higher level assembly to determine whether interchangeability has been maintained at this next higher level. In the example it has, so the revision level of part number X is changed from 0 to 1. If there had been 20 levels in the bill of material the same decision sequence would have taken place on all. Whenever a change occurs at one level, the next higher level must be examined to determine whether a part number or a revision level change is indicated.

In the example in Figure 92, the change notice is initiated to part number X. In addition, the drawing for part X must be revised, the detached parts list for X must be changed, a release notice for new assembly B and part 4 must be issued, and a new detached parts list for B prepared and issued.

Later we shall discuss a most important feature of a good bill of material system, *effectivity,* which keeps engineering and manufacturing documentation and the components of the physical product synchronized.

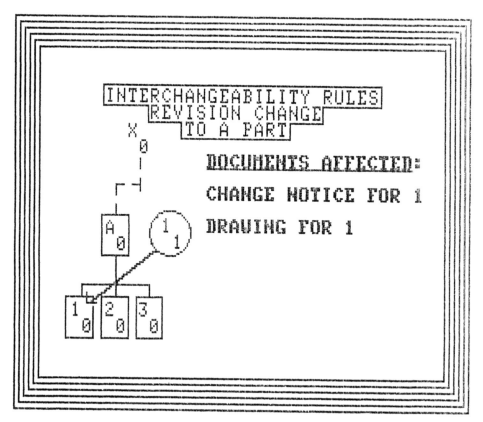

Figure 90 A simple change to a piece part which does not require a part number change is relatively simple.

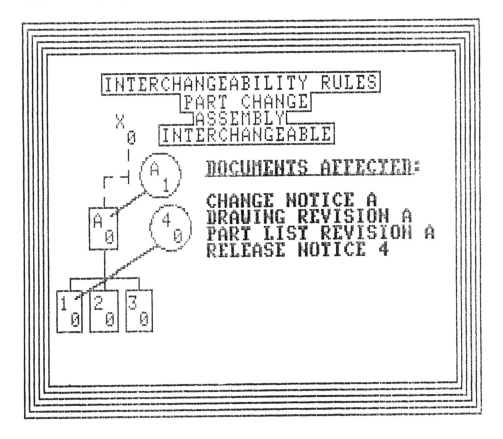

Figure 91 Replacement of one part for another requires investigating the influence of the change on the next level of the bill of material. The number of documents affected are higher.

BILLS OF MATERIAL

So far we have used the terms "structure," "level," "bill of material," and "parts list" or "assembly" rather loosely. Let's clarify some of these phrases. Figure 93 is a stylized picture of a *structure*. It represents the interrelationship of parts and assemblies in a *bill of material*. The circled section of the product structure represents one *level* in a chain of the bill of material, and represents an *assembly*. This assembly is documented utilizing a *parts list*. Documenting the relationship of all parts and each assembly and subassembly one to another results in the bill of material.

DATA PROCESSING TECHNIQUES

When converting a product identification system from a manual to a computer system, economies of speed, reduction of documentation effort, and integration of the system into manufacturing systems become possible. Additional forms which replace manual documentation methods are required. For example, the part number log is

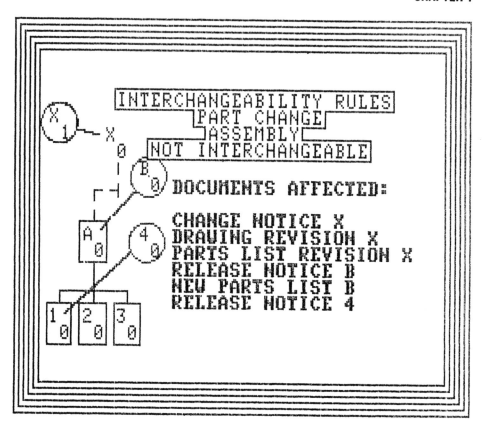

Figure 92 When the replacement of one part for another requires a change to the next level part number, documents affected become more complex.

replaced with a *release notice,* and the manual parts list is replaced with a parts list input form. Computer files are required and are thought of as data bases which contain predefined data increments.

For many years the input forms, the paper work, and the manual input "batch" methods implied above actually existed. Today, more and more systems are on-line, real time, eliminating much of the former bureaucratic paper work. As we continue, however, it will be clearer to describe activities of a system in terms of data on a piece of paper and its movement.

General practices have now made rather common the use of the words "item master," and "product structure" to describe the data base and system used by engineering for a bill of material system and referred to as the *engineering control model.* The item master contains a good deal of additional information utilized by manufacturing and other departments. Slowly, but surely, the item master is becoming considered as a significant portion of a company data base. We will look at some of the ingredients of the mechanized bill of material system.

Data Increments (Elements)

Each predefined piece of information required for a bill of material is called a *data increment*. Examples are part number, part description, unit of measure, part type code, lead time, etc.

Data Base

As data increments are defined for a system and then accumulated, the accumulation is called a *data base*. In most data base systems at least one key (such as a part number), is used to retrieve data as required. The use of a data-base technique is substantially superior to the old method of file-oriented banks of data.

As companies rely more and more on the engineering data base, the number of data increments grow. It is not uncommon to find 200–300 individual data increments available. Figure 94 lists a few of them, for both the item master and the product struc-

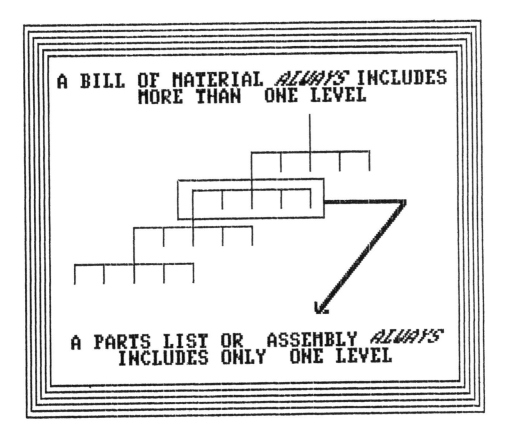

Figure 93   For clarity, it is important to be able to distinguish the difference between a parts list and a bill of material. Each document has its own requirements and is used for different purposes.

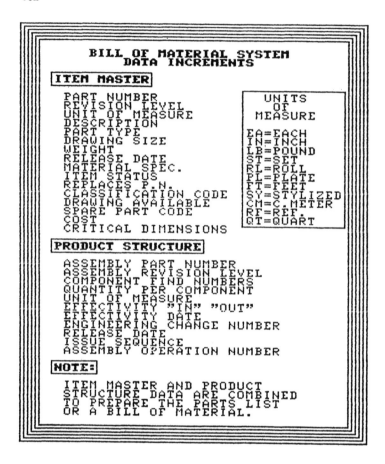

Figure 94 A mechanized bill of material has two associated "files" in the data base, the item master, and the product structure. Merging these two files produces many documents without the necessity of creating redundant data. Some of the usual data increments found in the system are shown.

ture. These data increments in a bill of material system, are normally divided into two segments of the data base: (1) item master and (2) the product structure files.

Item Master File

The *item master file* contains all data increments required for each part number in the system. The release notice is closely associated to this file and is used by engineering and manufacturing when a new design is released.

Release Notice

When a drawing is completed or a new part number has been issued, the *release notice* is the document which is used to inform all departments affected. It is the form used to capture data required by the system to prepare data found on parts lists and bills of materials. It is also used as a method for notifying departments other than engineering of the existence of a new design. All part numbers, when issued, require a release notice whether the design be a piece part, an assembly, or a purchased part or raw material.

The release notice is designed as a multiuse form. Upon release of a design by drafting, the appropriate raw material size, quantity, unit of measure, and part number of the raw material, when applicable, will be added by manufacturing to the document. In addition, for special conditions as on a "prerelease," additional data for the prerelease requirements may be inserted. Figure 95 is a typical release notice.

In summary, the release notice is utilized by the system to perform the following functions:

Release a new part number

Change an old design

Obsolete an old design

Inactivate an old design

Prerelease a new design

Figure 96 shows a general flow of data from the drafter to the computer utilizing a cathode ray tube (CRT) to enter the data.

Structure File

The *structure file* maintains the part assembly relationships and associated data uniquely required for an assembly. This is also often called the parent–component relationship. A component in an assembly can also be used on many other assemblies. Thus, the structure file maintains the component to parent relationship and the quantity of the components required for that relationship. The structure file is maintained using a parts list input form.

Parts List Input Form

The *parts list input* form is actually an abbreviated method for entering a parts list into the computer. Since all other data needed for a printed parts list is already stored in the item master file, only the parent–component relationships, quantities, find numbers, and other data associated with a product structure need to be entered into the system. When a parts list or bill of material is printed from the computer, the computer extracts the necessary additional data from the item master and combines it with the product structure data. Thus, the system eliminates the need for rewriting redundant data.

Figure 97 shows a typical parts list input. Figure 98 shows a general flow of data from the drafter to the computer utilizing a CRT. A typical parts list output is shown in Figure 99. Many additional reports are possible from this system such as:

Product "used on"

Indented bill of material

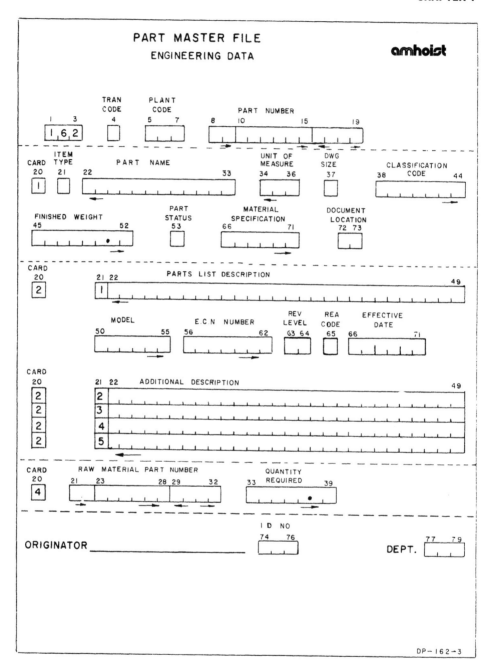

Figure 95　This form illustrates a release notice used with a mechanized bill of material system. If converted to an on-line and interactive system, the same form would appear in the form of a menu on the screen. The release notice is used to develop the item master file. (Courtesy of American Hoist and Derrick, St. Paul, Minn.)

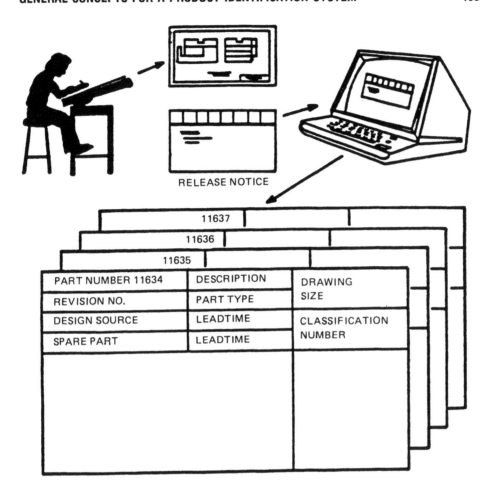

RELEASE NOTICE

11637		
11636		
11635		

PART NUMBER 11634	DESCRIPTION	DRAWING
REVISION NO.	PART TYPE	SIZE
DESIGN SOURCE	LEADTIME	CLASSIFICATION
SPARE PART	LEADTIME	NUMBER

Figure 96 The basic flow and data of the item master is shown here.

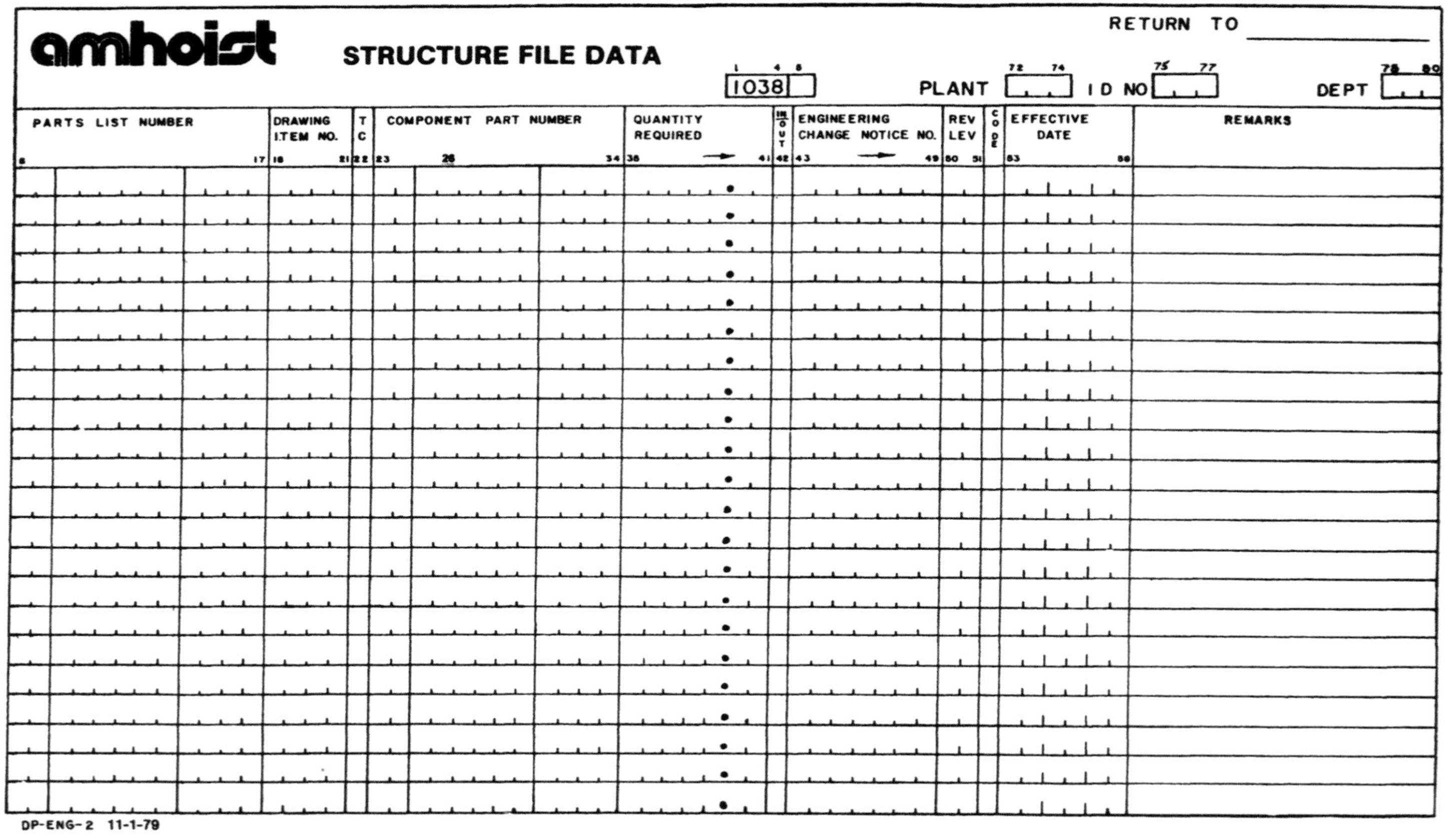
amhoist

STRUCTURE FILE DATA

RETURN TO

1 4 8

1038

PLANT 72 74

I D NO 75 77

DEPT 78 80

PARTS LIST NUMBER 8 17	DRAWING ITEM NO. 18 21	T C 22	COMPONENT PART NUMBER 23 26 34	QUANTITY REQUIRED 35 41	IN OUT 42	ENGINEERING CHANGE NOTICE NO. 43 49	REV LEV 50 51	CODE	EFFECTIVE DATE 53 58	REMARKS

DP-ENG-2 11-1-79

Figure 97 This form illustrates the parts list data required to develop the product structure file. An on-line system would use this as a menu to enter the data on the screen. (Courtesy of American Hoist and Derrick, St. Paul, Minn.)

Figure 98   The flow of data for the product structure is shown here together with the results, including the combining of data from the item master and product structure files to eliminate redundancy.

```
ANAREN MICROWVE                                                          AMPLOO

OPER MAH              A N A R E N   P A R T S   L I S T    DATE 12-05-82  TIME 10.24.30

AMI P/N  25248-G001     CLASS  -176C-      DESCRIPTION  PCB ASSY,THRESHOLD DET    PAGE 01 OF 00

        REV  05   DATE  12-08-82   ECN  123456   BY  234    REVISED & REDRAWN AS DEV DOCUMENT
        REV  04   DATE  10-15-82   ECN  234567   BY  128    RELEASED AT CLASS B
        REV  03   DATE  06-17-82   ECN  778899   BY  064    UPDATE TO PWB
        REV  02   DATE  05-14-82   ECN  123211   BY  272    DELETE TRANSORB
        REV  01   DATE  05-13-82   ECN  543210   BY  345    CHANGE U1
        REV  00   DATE  03-30-82   ECN  765432   BY  076    ORIGINAL REVISION

FIND PART NO.    CLASS  DESCRIPTION              QUANTITY  UM IT  REFERENCE DES. / NOTES   DATE I/O

0001 25247-0001 -175D  PCB,THRESHOLD DET           1.000  EA  4                             120882 I
0002 25180-0005 -210B  CONN,STR,D-PCB,SUBMINI      1.000  EA  4  J1                         120882 I
0003 19133-0001 -210A  CONN SMB                    6.000  EA  4  J5,J6,J7,J8,J9,J10          120882 I
0004 23370-0003 -313A  IC,LH0033C BUFFER AMP       1.000  EA  4  U1                         120882 I
0005 24785-0003 -313A  IC,26LS32 LINE REC          1.000  EA  4  U2                         120882 I
0006 17601-0003 -313A  IC,54**123J                 5.000  EA  4  U3,U7,U8,U20,U21           120882 I
0007 20443-0003 -313A  INTEG CKT 54S112 BEB883/B   2.000  EA  4  U4,U9                      120882 I
0008 13516-0003 -313A  IC,5400,QUADR 2 INP NAND    2.000  EA  4  U5,U13                     120882 I
0009 25491-0003 -313A  IC,54*S260J DUAL 5IN POS    1.000  EA  4  U6                         120882 I
0010 24755-0003 -313A  IC,54*S74 DUAL D            1.000  EA  4  U10                        120882 I
0011 24786-0003 -313A  IC,26LS31 LINE DRVR         1.000  EA  4  U11                        120882 I
0012 24154-0003 -313A  IC,54*S30J INPUT NAND       1.000  EA  4  U12                        120882 I
0013 25204-0003 -313A  IC,SE521F COMP HI SPEED     2.000  EA  4  U14,U15                    120882 I
0014 19188-0002 -313A  IC,54*S04J,HEX INVERTER     1.000  EA  4  U16                        120882 I
0015 10777-0002 -313A  AMPL,OPNL,UA741             3.000  EA  4  U17,U18,U19                120882 I
0016 20864-0001 -313A  DC RESTORER MODULE          1.000  EA  2  U22                        120882 I
0017 12069-0003 -346A  DIO-SIG-P5-1N914-75V        2.000  EA  4  CR1,CR2                    120882 I
0018 13267-0001 -538A  DIO-ZEN-1N827-8.0V-400      2.000  EA  4  VR1,VR2                    120882 I
0019 12072-0003 -538A  DIO-ZEN-P5-1N753-6.2V-400   2.000  EA  4  VR9,VR10                   120882 I
0020 10666-0102 -470A  RES FIXED FILM 1/4W 2%      2.000  EA  4  R1,R10                     120882 I
0021 10665-0102 -470A  RES FIXED FILM 1/8W 2%      8.000  EA  4  R2,R3,R4,R5,R6,R7,R8,R11   120882 I
0022 10665-0472 -470A  RES FIXED FILM 1/8W 2%     20.000  EA  4  R9,R24,R25,R26,R27,R28,    120882 I
                                                                 R29,R30,R31,R32,R33,R34,
                                                                 R35,R36,R37,R38,R39,R40,
                                                                 R41,R91
0023 10665-0242 -470A  RES FIXED FILM 1/8W 2%      2.000  EA  4  R12,R13                    120882 I
0024 10665-0332 -470A  RES FIXED FILM 1/8W 2%     11.000  EA  4  R14,R15,R16,R17,R18,R19,   120882 I
                                                                 R20,R21,R22,R23,R93
0025 10665-0512 -470A  RES FIXED FILM 1/8W 2%     10.000  EA  4  R42,R43,R44,R45,R46,R47,   120882 I
                                                                 R48,R49,R50,R73
0026 10665-0103 -470A  RES FIXED FILM 1/8W 2%      4.000  EA  4  R51,R52,R53,R54            120882 I
0027 10822-0624 -470A  RES GLASS EPOXE 1/8W 2%     4.000  EA  4  R55,R56,R57,R58            120882 I
0028 10665-0100 -470A  RES FIXED FILM 1/8W 2%     10.000  EA  4  R59,R60,R61,R62,R63,R65,   120882 I
                                                                 R66,R67,R68,R94
0029 10665-0221 -470A  RES FIXED FILM 1/8W 2%      2.000  EA  4  R64,R92                    120882 I
0030 10665-0101 -470A  RES FIXED FILM 1/8W 2%      4.000  EA  4  R69,R70,R71,R72            120882 I
0031 10666-0101 -470A  RES FIXED FILM 1/4W 2%      3.000  EA  4  R74,R75,R76                120882 I
0032 12864-0502 -475A  POT 3262W 5K                1.000  EA  4  R77                        120882 I
0033 12778-0103 -475A  POT CERMET 10K             13.000  EA  4  R78,R79,R80,R81,R82,R83,   120882 I
                                                                 R84,R85,R86,R87,R88,R89,
                                                                 R90
0034 12787-0680 -470A  RES FIXED FILM              2.000  EA  4  R101,R102                  120882 I
0035 10880-0104 -150A  CAP CK06 0.1 UF            29.000  EA  4  C1,C2,C3,C4,C5,C6,C7,C8,   120882 I
                                                                 C9,C10,C11,C12,C13,C14,
                                                                 C15,C16,C17,C18,C19,C20,
                                                                 C21,C22,C23,C24,C25,C26,
```

Figure 99 The parts list shown here is one output from a mechanized bill of material system. Special data appears on this document for PC board documentation, a sign of the flexibility of such a system. (Courtesy of Anaren Microwave, Inc., Syracuse, NY.)

Picking list

Next level "used on"

Figure 100 is an example of these reports. Divided between bill of material, used-on, file extraction, and bill of material "explosions," the reports are powerful byproducts of the system.

SPECIAL CONSIDERATIONS

Prerelease

Due either to a long lead time, or when a part, assembly or product is being manufactured as a *prototype,* new parts and assemblies can be *prereleased.* When a part is in the prerelease stage, the revision level must be other than standard, such as alpha versus numerical. The item master file isused to maintain this status. Less rigorous change procedures need to be followed and limited production or procurement is authorized only.

Figure 101 pictures three levels of documentation control. If used judiciously the maximum utilization of a bill of material will occur. Let us look at these levels.

Level A: Class A Documentation

Class A documentation is fully controlled documentation. Formatted drawings and parts lists are released, and changes are processed through a change control board. Class A documentation is used for production.

Level B: Class B Documentation

Class B documentation is also called prerelease documentation. Sketches are used in lieu of formatted drawings, but are controlled through an engineering change process. Interchangeability of parts rules are recognized. Class B documentation is utilized for building a predefined but limited number of production prototypes.

Level C: Class C Documentation

Class C documentation is used for research and development of complex systems. It provides a tool for early identification of parts and assemblies and the building of a product structure for manufacturing. There are no formal revision control or requirement for sketches or formatted drawings. With few or no exceptions, no parts or materials may be ordered for parts having a class C status.

Figure 101 diagrams three parts lists. These indicate the various stages an assembly might pass through. In the first stage, assembly part number A is composed of parts 1, 2, 3, and 4. Parts 1 and 3, are in a class B status, having revision levels of (01) and (04), respectively. Part 2 is in class A status having a revision level of (B). Part 4 and assembly A are in class C status, having no revision level. Thus, a class C assembly may contain a variety of parts with different class status.

In the second stage, the assembly has been converted to a prerelease status (01). Notice that all parts within the assembly are now either prereleased or released. The

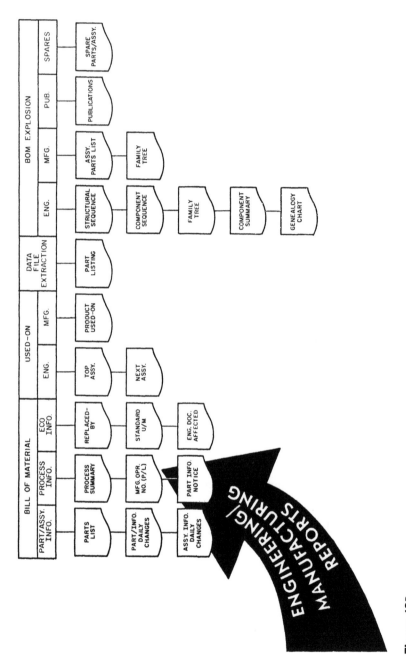

Figure 100

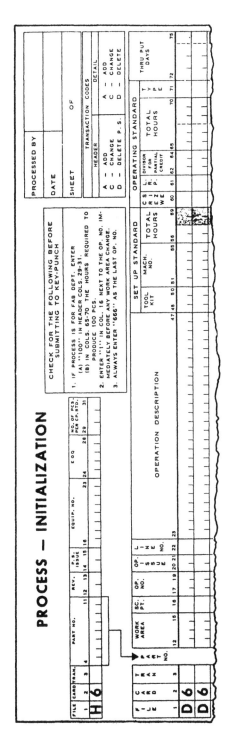

Figure 100 Many different forms of data output can be generated for a bill of material system. Thus, the bill of material becomes the "spine" of all associated systems.

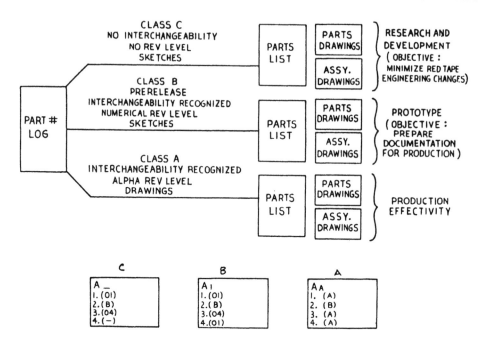

Figure 101 The use of three separate levels of documentation in engineering provides versatility and flexibility.

implied rule states that no part of an assembly can be at a lower level of documentation than the assembly.

The third stage shows the assembly A as a released document. All parts within the assembly have also been released. The complete design is ready for production. However, during the first and second stages, parts and materials were allowed to be ordered, and the product structure was available to both engineering and manufacturing.

Figure 102 shows a powerful tool for use with class C documentation. The document shown is an indented bill of material for all parts in a product backed off by lead times. Even though many or all parts are yet to be designed and drawings prepared, this "lead time bill of material" shows engineering, purchasing, and manufacturing the critical parts requirement. In addition, if engineering design lead time is added to all parts not yet released, the engineering design and drafting schedule may be prepared.

Raw Material

Each part made from raw material requires a single line parts list showing the part number of the material to be used and the quantity of this material required to make one part. Since this relationship is the responsibility of industrial engineering, the raw material size may be changed at any time without an engineering change notice, revision level, or part number change. Figure 103 shows one way to describe the type of raw material to be used. It is most important that the proper units of measure be ap-

LEAD TIME CHART

PART NUMBER	DESCRIPTION	LEAD TIME	M\P	LC	6/25 365	8/23 330	10/05 300	11/16 270	1/07 240	2/18 210	4/01 180	5/13 150	6/27 120	8/16 90	9/28 60	11/09 30	12/23 0
18347-G001	02*0755 DI	30	M	0													XXXXXXX
18343-G001	CASE	30	M	1												XXXXXX	
20147-0009	AL,6061-T6 .190	25	M	2											XXXXX		
18344-0001	GND PLN	25	M	2											XXXXX		
10739-1300	PTFE,GL LAM,TEF	45	P	2										XXXXXXXXX			
18344-0002	GND PLN	25	P	2											XXXXX		
10739-1300	PTFE,GL LAM,TEF	45	P	2										XXXXXXXXX			
10739-1310	PTFE,GL LAM,TEF	45	P	2										XXXXXXXXX			
18344-0003	GND PLN	25	M	1											XXXXX		
10739-1048	PTFE,GL LAM,TEF	45	P	2										XXXXXXXXX			
18344-0004	GND PLN	25	P	1											XXXXX		
10739-1300	PTFE,GL LAM,TEF	45	P	2										XXXXXXXXX			
18344-0005	GND PLN	25	M	1											XXXXX		
10739-1300	PTFE,GL LAM,TEF	45	P	2										XXXXXXXXX			
18345-G001	ASSY CKT	19	M	2											XXXX		
17834-0001	CKT	20	M	3									XXXXXXXXX				
10739-1053	PTFE,GL LAM,TEF	45	P	2								XXXXXXXXXXX					
10868-0100	CAP CHIP 10 PF	70	P	2								XXXXXXXXXX					
10838-0500	RES,CHIP,100 MW	50	P	2								XXXXXXXXXX					
10838-0101	RES,CHIP,100 MW	50	P	2								XXXXXXXXXX					
10838-0221	RES,CHIP,100 MW	50	P	2								XXXXXXXXX					
10838-0431	RES,CHIP,100 MW	50	P	2								XXXXXXXXX					
10327-0002	FEED THRU STOCK	10	M	2												XX	
20156-0002	BRASS SHIM BOXE	25	P	3								XXXX					
16265-0001	CONNECTOR	100	P	1							XXXXXXXXXXXXXXXXXXXXXX						
12700-0003	CHANNEL INSERT	25	M	1											XXXXX		
14599-3310	PTFE GL LAM ROG	52	P	2							XXXXXXXXXXXXX						
12333-G005	FILTER ASSY 130	19	M	1											XXXX		
14601-0001	FILTER	70	P	2							XXXXXXXXXXXXXX						
14089-0001	WIRE,#34 AWG	55	P	1							XXXXXXXXXXXXX						
12831-0001	8-32 HEX NUT	40	P	1									XXXXXXXX				
14613-0001	DIO-SCH-P3-H616	51	P	1								XXXXXXXXX					
12069-0001	DIO-SIG-P5-1N91	40	P	1								XXXXXXXXX					
10822-0203	RESISTOR COMP 1	50	P	1								XXXXXXXXXX					
12615-0001	RFI GASKET D ST	50	P	1												X	
17371-0008	WASH CHROM MF 1	5	M	1										XXXXXXX			
12773-0001	COND ELAST SHEE	35	P	2										XXXXXXX			
12939-0001	WIRE 30 AWG BRO	35	P	1											XX		
10327-0003	FEED THRU STOCK	10	M	1										XXXX			
20156-0002	BRASS SHIM BOXE	25	P	2							XXXXXXXX						
10630-2004	2-56X5/16 PH FL	50	P	1							XXXXXXXX						
10633-0002	2-56X3/16 PH RN	50	P	1							XXXXXXXXX						
11521-0077	#2 WASHER SPLIT	50	P	1							XXXXXXXXX						
17105-0001	GROUND LUG	35	P	1								XXXXXXX					
16123-G001	PRIMER & CATALY	35	P	1								XXXXXXX					
16221-G001	PAINT,BLUE	35	P	1								XXXXXXX					
13734-0001	TAPE,KAPTON	35	P	1								XXXXXXX					
13828-0001	EASTMAN 910	35	P	1								XXXXXXX					
17960-0001	63/37 SOLDER	45	P	1								XXXXXXXXX					
13382-0001	SILVER EPOXY	39	P	1								XXXXXXX					
18346-XXXX	ASSY DWG	4	M	1												X	

Figure 102 When properly developed the bill of material system will provide a lead-time chart for designs which have yet to be prepared. Thus, design and drafting schedules become a by-product of the system.

Descriptive nomenclature	Material size or key section dimension	Material description (balance of dimensions)	Description
Bar, sheet, strip, and plate	Thickness	× Width × length	Description
Squares and round-cornered squares	Thickness	× Length	Description
Hexagon	Across flats	× Length	Description
Rounds and half-rounds, wire, wire rope, drill rod	Diameter	× Length	Description
Ovals and half ovals	Width	× Thickness × length	Description
Angle	Larger leg	× Smaller leg × thickness × length	Description
Angle, bar	Larger leg	× Smaller leg × wt/foot × length	Description
Channel, Bar size	Depth	× Flange × web thickness × length	Description
Channel, structural beams	Depth	× Flange × web thickness × wt/foot × length	Description
Tee, bar size	Flange	× Depth × thickness × length	Description
Tee, structural	Flange	× Depth × thickness × wt/foot × length	Description
Tubing, round	Outside dia.	× Wall thickness × length	Description
	Inside dia.	× Wall thickness × length	
	Outside dia.	× Wall thickness × length	
Tubing, square	Outside dimen.	× Wall × length	Description
Tubing, rectangular	Larger outside dimension	× Smaller outside dimen. × wall × length	Description
Pipe	Nominal dia.	× length	Description

Figure 103 Raw material titles and data are easier to mechanize if organized in a standard manner.

plied to the "quantity" field of the parts list. Otherwise, serious errors in the procurement of raw material may occur. Once incorporated in the bill of material, however, a rapid means to compute total raw material requirements becomes available. Price changes of raw material are also expedited with this technique.

Effectivity (Date When Change Is Effective)

Returning to the product structure, let us discuss an aspect of its power and compatibility for both engineering and manufacturing documentation. For many years these departments had apparent conflicting needs in a bill of material system. Often an engineer would make an engineering change, removing, for example, one part from an assembly, and replacing it with another. To engineers the change became effective at the moment of alteration. However, to manufacturing, the change might well be impractical or uneconomical in manufacturing until weeks or even months had elapsed. This change lag is both logical and practical. Part inventories or purchased-part lead time created the need for two bills of material, one for engineering showing current design requirements, and one for manufacturing showing current production requirements. A mechanized bill of material, properly designed, eliminates this redundancy.

Figure 104 shows an example of effectivity. The top two lines of the parts list show part number 00838300 as removed, or "out," and part number 00838301 replacing

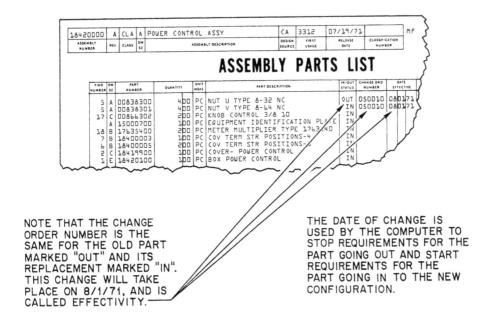

Figure 104 A mechanized bill of material system provides versatility by providing a means to document two or more configurations of an assembly using "effectivity." Thus, both engineering and manufacturing may use the same system.

it, or "in." Attached to this transaction is the date of change, 08/01/71, and the change order number 050010 which may be used to find the reason for the change.

This allows two configurations of one parts list to exist at the same time. The data may be used in manufacturing resource planning to stop the acquisition of the old part and initiate the procurement of the new part based on lead times.

Order-picking lists will also reflect the period of time when either the old or the new part is required for manufacturing. Thus, compatibility has been created, and the former redundancy and cost of two bills of material is eliminated. Needless to say, the cost accounting department reflects more accurate costs for the product.

Stylized Drawings

Frequently, several groups of parts may be put together to make different assemblies, but can still use identical drawings. To eliminate the need for drawing assembly drawings over and over again, use stylized drawings. Figure 105 is such a drawing, including the rules which must be followed to use it adequately. (Similar, but less formal, drawings are familiarly called a *reference* drawing.) The stylized drawing's part number always appears on the parts list for the parts to be assembled. The quantity column, however, is left blank or "zero."

The series of parts lists, A, B, C, D, E, F, and G each contain a list of different parts, but have identical find numbers. The stylized drawing part number 123456 appears on each parts list to identify it as the drawing which will be used to assist in the assembly procedure.

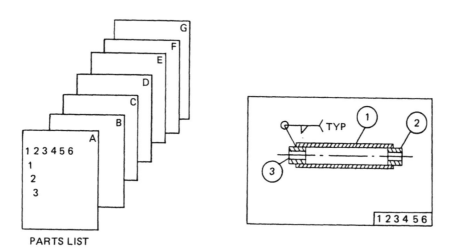

PARTS LIST

1. All "balloon" numbers shown on a drawing must have a corresponding piece part identified on *each* bill of material.

2. No stylized drawing may have a parts list on the drawing.

3. The part type code for a stylized drawing will be "s" for stylized.

4. When using a stylized drawing on a parts list "sty" will be entered in the quantity field.

Figure 105 Use of stylized drawings eliminates the need for many repetitious assembly drawings.

Engineering Change Review "Board"

Engineering change procedures are always expensive and often complicated. Unilateral changes by designers or drafters can be catastrophic. Success has been achieved by assigning people from engineering, inventory control, industrial engineering, and plant operations to an engineering change board. This board reviews the complex or potentially expensive changes to come to a consensus on when and if the change should be made.

Many companies have horror stories to relate on the degree of red tape surrounding an engineering change and the host of persons authorized to sign approval of the change. It is not uncommon to find a change taking as long as six months to be approved. This is intolerable.

A study must be made to eliminate the need for signatures of persons who may be only on the list for status. In addition, the change notice should be considered "electronic mail" so that those involved in authorizing it may sign it in parallel, not sequentially. Perhaps a saga of an engineering change, taken directly from the user of such a system will dramatize the problems.

An Engineering Change—Or The Kibuki Dance

1. A technician finds a problem on a P.C. board and explains to engineer in charge.
2. Engineer pulls capacitor out of junk drawer which has the correct value.
3. Technician verifies fix and tells engineer.
4. Engineer begins writing an engineering change notice (ECN).
5. Engineer asks and searches for part number to see if it exists.
6. Engineer searches through poor and incomplete parts lists.
7. Engineer searches through documentation parts notebook.
8. Engineer realizes that part is probably not in the company system and has to create a part number.
9. Engineer reproduces all part information and fills out check list.
10. Repeat "9" until drafting is satisfied.
11. Present new parts list change and complete ordering information (vendor, delivery dates, etc.), and explain to production control and/or purchasing. (Get signature.)
12. Call back to make sure part is ordered.
13. Find out why Pat didn't get into the computer data base and get it done.
14. Update the parts list.
15. Sign the ECN.
16. Find the supervisor and explain all is ok. (Get signature.)
17. Submit ECN.
18. Repeat "17" until drafting is satisfied.
19. Correct any mistakes when drafting is finished—(another ECN?)
20. Correct new part schematics when drafting is finished.

Is it worth the effort to prepare and maintain a good bill of material system? If a company desires an integrated data base to perform rapidly and accurately in today's business climate, it becomes mandatory. The engineering bill of material is the core, the focus, of every other system. In addition, there are some sound reasons to recognize that a mechanized bill of material is more economical than one prepared manually. Some of these reasons appear in Figure 106.

Indeed, Figure 107 shows how an on-line, integrated system uses a common data base. The power of such a system is an order of magnitude greater than manual and/or noninteractive systems.

The bill of material, when tied to the master schedule for manufacturing, becomes the key document to prepare total material requirements for the company. Figure 108 demonstrates this event.

Utilizing manufacturing resource planning, this material may also be time phased for delivery to minimize inventory.

Implementation

It is important to be aware of the amount of work necessary to implement a bill of material system which has integrity and also the respect of users. Figure 109 shows an example of implementation in terms of time and activities developed by one company. Notice that the first 6 months were used to prepare an adequate manual bill of material system before converting to a mechanized system. The hub of the manual system became the part number requirements. (The company had a 45-digit part num-

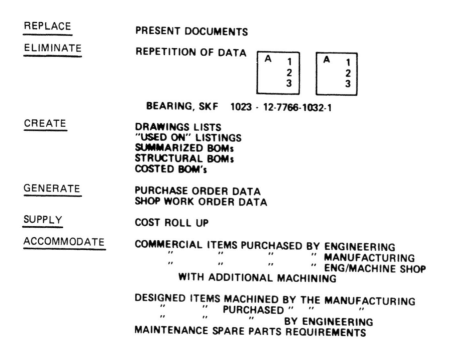

Figure 106 Some typical reasons for mechanizing the bill of material system are shown here.

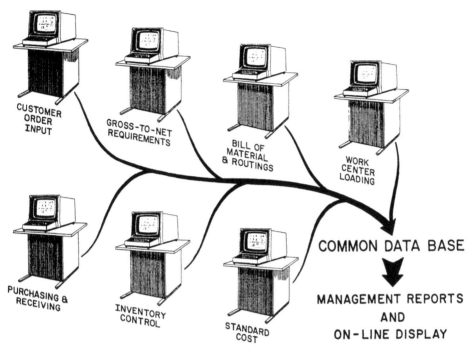

Figure 107 The power of a bill of material system can be felt if it is considered to be the pipeline through which data flows and which can then be tapped by all involved personnel. If on-line, the data becomes timely, accurate, and readily accessible.

ber, so conversion to a new one became almost mandatory.) The hub of the mechanized system became the item master and product structure files.

The following list of policies, standards and procedures were prepared for this system, and a record 800 people were instructed on each of them during the 18 months of design and implementation. The system is working most effectively.

ENGINEERING STANDARDS

Drafting standards
    Drawing content
    Drawing standards
    Parts list
Design standards
    Equipment structure
    Modular design
    Design retrieval
    Quality standards

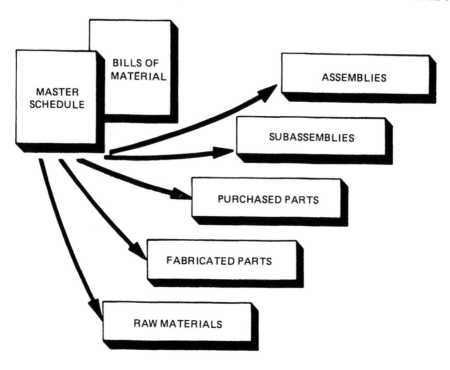

Figure 108 If product requirements in the form of a master schedule is prepared, and the parts and assemblies required are documented in a bill of material, multiplying one by the other generates total requirements of all raw materials, parts, and assemblies.

Configuration control standards

Engineering change notice

Interchangeability

Part number/revision level

Effectivity of change

Equipment configuration control

Equipment documentation standards

Equipment specifications

Deviation authorization

Field changes

Engineering release

Serialization

Spare and repair parts control

Special equipment documentation

Material specifications

Units of measure

Microfilm

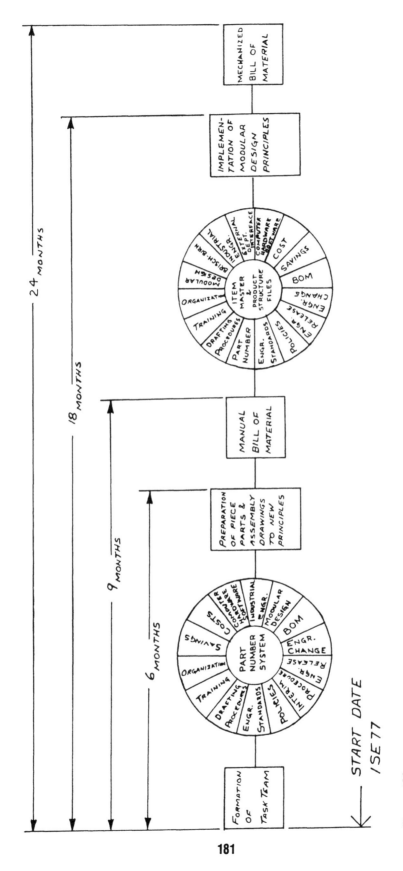

Figure 109 Mechanization of a bill of material system requires significant planning and scheduling to avoid interrupting working systems. Meticulous care must be taken to implement a system in an orderly, logical sequence.

181

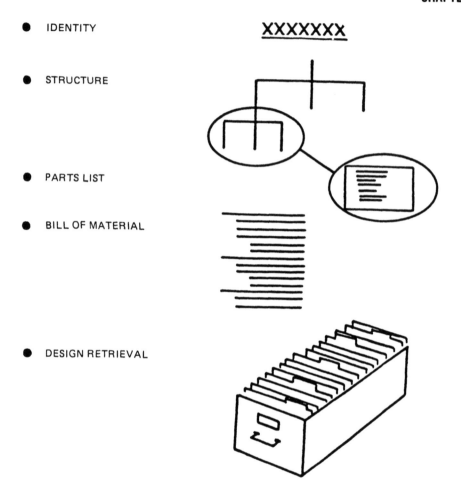

- IDENTITY
- STRUCTURE
- PARTS LIST
- BILL OF MATERIAL
- DESIGN RETRIEVAL

Figure 110   The minimum requirements for implementing a successful bill of material include these items.

Bill of material (manual)
Bill of material (mechanized)

In the same manner, other segments of the system were defined, and activities defined. Some of the segments follow:

Part number system
Industrial engineering system
Computer hardware
External department interfaces
Interim system
Training

Costs

Savings

Policies required

Definitions required

All these general categories of activities were shown on a program evaluation and review technique (PERT) chart.

The requirements shown in Figure 110 were met, that is creating an identity or part number, preparing product structures, generating parts lists, creating a bill of material and implementing a design retrieval system.

CONCLUSION

The principles, rules, definitions, and systems indicated in this general concepts section cover the major segments of a modern engineering documentation system. Interrelationships between manufacturing and accounting systems have not been described or indicated.

In order to describe the basic systems for an engineering bill of material, an online system has not been emphasized. This does not mean it is not important. Quite the contrary, the only completely adequate way to make a bill of material and its associated systems effective is through on-line update and maintenance. Such paper work as release notices, parts list input, and the engineering change notice become *menus* on the screen of a cathode ray tube.

If the basic concepts described here are kept intact most methods of implementation will work effectively.

PRODUCT STRUCTURING

8

We have devoted considerable text to the ingredients of an engineering documentation system. Each of these ingredients are important to the integrity of the documentation and to the flow of data. Now that this data has been indicated, the next step begins the data movement through the use of a bill of material system and its associated parts lists and item master data.

It would be wasteful to spend time and money to develop a system with high integrity only to find it useless because the data is inaccurate. This problem often occurs. The bill of material is structured by engineering and formatted as they think manufacturing needs it. If, on the other hand, manufacturing must change it to reflect actual manufacturing and assembly methods, a duplication of effort and even duplication of data on the computer may result. Thus, the recognition of the need for structuring the bill of material becomes important.

What do we mean by "structuring" a bill of material. How will this be done? How will proper documentation be maintained? This chapter will address these questions and set up guidelines to structure bills of material.

The first bill of material was erected in 2800 BC. As you can see in Figure 111, it had a large number of levels, its structure was complex, but the part relationships were relatively simple! It was simply stone, stone, stone, and more stone!

It will help us, however, to understand this principle of a pyramidal structure, because, as we will soon see, a product structure and its corresponding bill of material follow this model quite closely.

Strangely enough the structure we want to discuss looks like a skeleton of the pyramids shown in Figure 112, however, the bill of material structure as we know it today is far more complex in many products, and may have 20 or 30 levels. We also refer to these structures as *bills of materials*. We say they are comprised of *assemblies* and *components*. We speak of the *parent-component relationships*. Let's look more closely at these structures and see what they can do for us. Figure 112 shows a transitional picture from the pyramid to a bill of material product structure.

Studying the simple diagram in Figure 113, let's define what it represents:

Product structure   A graphical or pictorial relationship showing the hierarchical

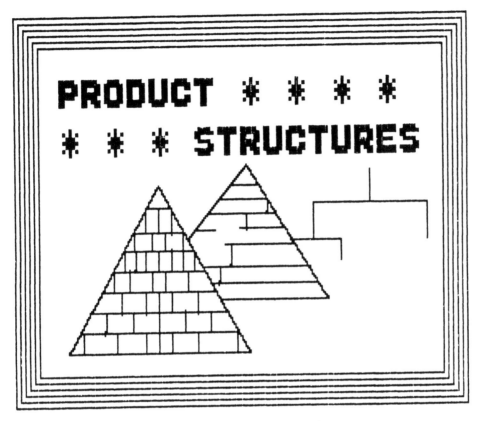

Figure 111   The concept of product structures are not new!

relationships and combination of parts and assemblies in a product. Furthermore, these relationships portray the manufacturing assemblies and *levels* of each part and assembly.

This relationship has other names. It has also been called a geneology chart, a family tree, and a Christmas tree.

As a basic technique and tool, the product structure should be thought of as two distinct types; (1) the general and (2) the specific. The general product, shown at the top of Figure 114, should be considered the means to "diagram" how a product fits together in parts and assemblies. The specific product structure should be considered an actual example of a product.

In this example, the general product structure includes the names of the parts for a pump impeller. The product structure thus prepared also illustrates the assembly sequence. Below the general product structure appears a specific product structure with part numbers for actual parts defined precisely. Thus, the general assembly sequence and location of each part is first defined as a general condition, and then followed closely when a real product design is documented. Later we will indicate some of the rules tied to the specific product structures which make them the key tool for preparing a proper bill of material.

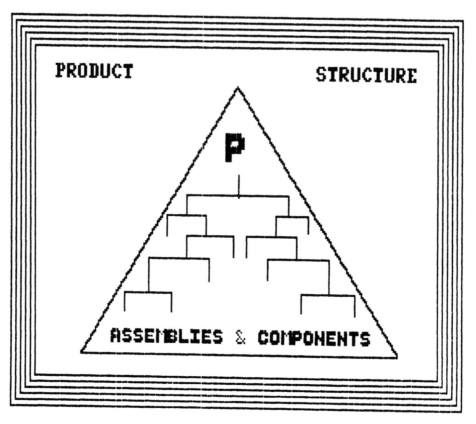

Figure 112 As products have become more complex, a simple listing of all parts required has become insufficient.

Figure 115 shows another product structure, in this case each leg of the structure has a part number. For simplicity, the letters and numbers shown here represent actual part numbers.

By definition, the product structure is that of a specific product, and represents an exact physical entity which can be found and identified in the manufacturing process. An assembly A, B, or C must physically exist in the product, and become work-in-process during manufacturing. Assembly B or E with components 1, 2, and 3, or 4, 5, and 6, must also exist.

To utilize a product structure in a practical way, it must be converted into a bill of material. Here we see the product structure converted into a form of bill of material known as an "indented," or "structural," or "structured" bill of material. As you can see, the conversion of a graphical representation to a listing of the structure starts by showing the product part number A on the top line and in the left-most position.

Product A is composed of assemblies B, C, D, and E, but B is composed of 1, 2, and 3, and E is composed of 4, 5, and 6. Each time we drop down a level in the product structure, we *indent* the line on which that part or assembly appears. Thus, at the end, we have a bill of material *structured* to show part-assembly relationships. If we rotate the indented bill of material counterclockwise 90 degrees, we show each level and the parts and assemblies appearing on each level. We have reproduced our graphical representation faithfully into a more useable media.

Several times we have referred to a bill of material. Each of you no doubt have used a parts list, a list of material, or a bill of material at some time. For a moment reflect on this idea, and see how you would define a bill of material if you had the opportunity. We will then present a definition with which you can compare yours, as shown in Figure 116.

Bill of material A complete record of all parts and assemblies required to make a specific machine or product. It is the paper counterpart of a physical entity.

Three important types of bills of material exist. A structural or indented bill of material has already been shown; a level-by-level bill of material is a series of parts lists for assemblies; a summarized bill of material is a list of the total parts required regardless of their locations in the structure.

As we begin to combine the attributes of a product structure with a bill of material, it is necessary to more fully illustrate the various levels found in the structure bill of material. In Figure 117, the top level of the bill of material is shown to be a product.

The next level is composed of assemblies. Often there are those who attempt to

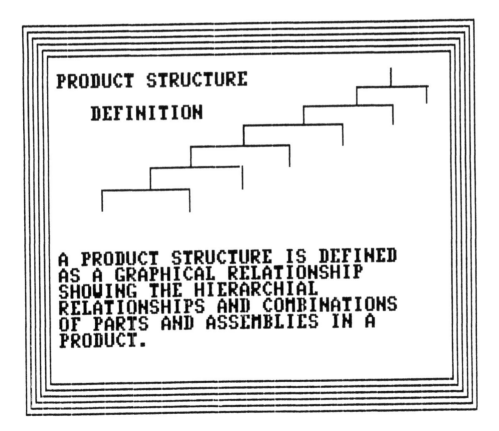

Figure 113 The definition of a product structure clarifies the reason for a bill of material.

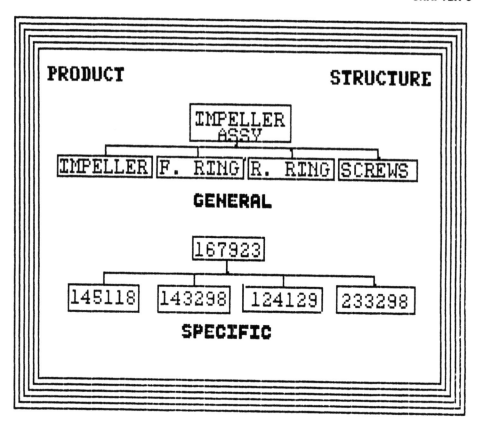

Figure 114 First a general product structure should be prepared before any specific product structure is documented.

further qualify assemblies into another category, subassemblies. When a product is fairly complex and has several levels, the use of assembly, and subassembly as a means for defining levels has no real meaning and may create confusion.

In any assembly four different types of parts can exist. Parts may exist at all levels mingled with assemblies. We have omitted alternates to avoid complicating the picture. The four types of parts are those which are fabricated, purchased, forgings, and castings. Through common practice, any purchased part, whether it be an electric motor or a bolt is called a part and has no lower level.

The finished forging and finished casting may create the need for additional levels under them, such as a rough forging or casting, and then an even lower level can be used to document the pattern. The fabricated part has an additional lower level, the raw material. It is thus evident that the lowest level in any bill of material is always raw material or purchased items, and can be used as a rule to check the bill of material for consistency.

Figure 118 is a kit of parts. A kit is a loose assemblage of parts which are not assembled but issued to manufacturing in a bag or tote box, or on a skid. In the example shown, the parts are such that it would be impractical to assemble them as a unit. The kit can be an important feature of a bill of material system, as a kit may be used to put two assemblies together.

A kit is useful for marshalling parts for large assemblies. For example, a set of gears, bearings, drums, and shafts with retaining rings could be moved as a unit and be assembled to a 10-ton cast machine body.

Figure 119 shows a simple assembly drawing and its corresponding parts list. We must use a detached parts list as such a list will no longer appear on the assembly drawing.

With the parts list detached from the assembly drawing, it is necessary to coordinate and correlate the two documents. As mentioned before, this coordination occurs using find numbers or bubble numbers on the assembly drawing. The drawing in the previous illustration has all the assembly data required, and also, each part of the assembly has been given find numbers. These numbers are nonsignificant and issued sequentially. The detached parts list shows each part number for parts needed in the assembly, the total quantity required, and the required find number for each part in the assembly drawing. Thus, a detached parts list and mechanized bills of material and product structures become practical and possible.

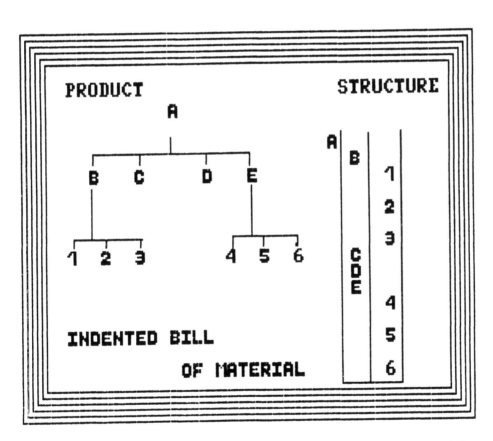

Figure 115 A product structure diagram can be converted to a printed document by following the rules shown here.

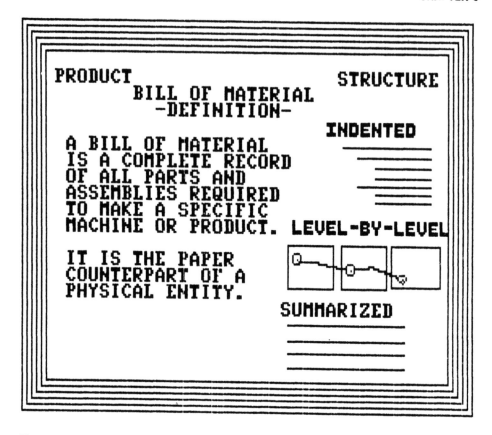

Figure 116 The definition of a bill of material shows how all-encompassing and diversified the system is.

After so much emphasis on the how, it is time to explain the why of product structures. Some of the important points are summarized on Figure 120.

During the past 20 years, many companies have developed bills of materials, so they are not a new idea, but some companies have been so enthusiastic, they have developed separate bills of material for each department, engineering, manufacturing, and accounting! Not only is this obviously expensive, it is confusing to design and build and costs differ. Purchasing a more modern data processing system does not eliminate the need for one bill of material. The product structure is the tool which will assist in this unification.

For many years, drawings were partially a picture, and both a picture and a list of material. The mechanization of the list portion soon brought to the attention of those who used them the obvious discrepancies between the drawing and the new mechanized list. Since a proverb of good systems is never to have two masters, elimination of redundancy became economical, but also important for accuracy.

Material requirements planning (MRP) was conceived as a tool to minimize stocking or purchasing materials prematurely. This theory, is dependent upon an accurate, time-sequenced bill of material. Should the required components not appear on their correct level of the bill of material, MRP effectiveness is diminished.

Products have become more complex, and have many variations. Later we will discuss modular design and a relatively new arrangement of bills of material to alleviate this problem. Complex products rely more and more on the proper structure of a bill of material. Job shops have found it the best way to program diverse one-of-a-kind designs.

Given that product structures are imperative for successful control of manufacturing, practical and logical procedures are needed to make them work, as indicated in Figure 121.

In virtually every manufacturing plant, many people are assembling, building, and designing using partially correct structures. Significant manufacturing changes which affected product structures years ago are followed even if the paper-work does not reflect them. If we will but merge all these disciplines into an ad hoc committee or task force, we will be successful.

Recognize that there is a sincere disagreement between departments as to how or why a product should be assembled in a certain manner. For example, assembly personnel may be assembling according to methods set in place by a foreman who has since retired. Stores choosing the material for the shop may be staging many levels of a multilevel bill of material. Industrial engineering may have long since discovered

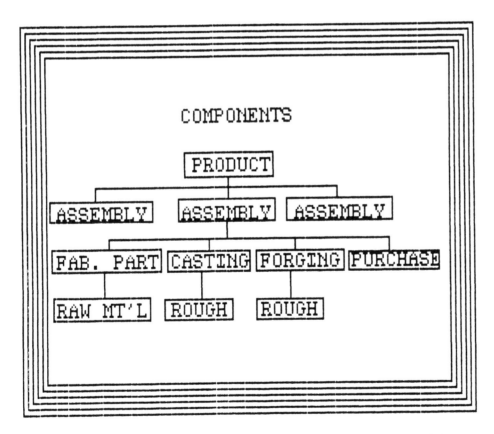

Figure 117 A general product structure can be quite complex, thus a picture of the part-assembly relationships is mandatory for clarity.

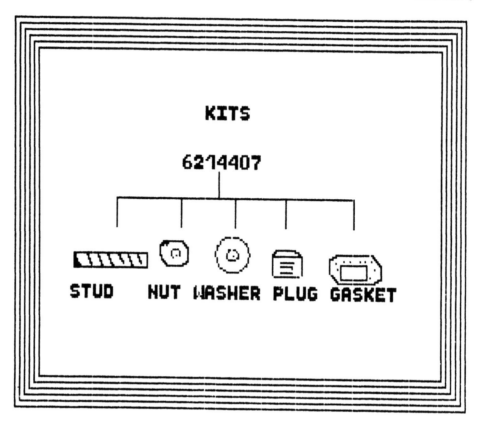

Figure 118 Special product structures as for a kit of parts are required and are perfectly correct.

a more economical sequence of assembly. Design engineering might disagree with the sequence and/or the method of assembly.

Product-by-product, the task force should agree on each phase of the product structure for it, and document it with a true drawing of the general product structure. This will require compromises between the various departments involved in the structuring exercise, but it is well worth it. In addition, policies and rules for the issue of new designs need to be made. It is not always possible to immediately change present bills of material. A planned transition or replacement will be required. It will be necessary to develop an education program for the hows and whys of the change. Last, before commencing, all must agree that there will be only one bill of material used by all departments in the entire company!

A permanent liason between design engineering and industrial or manufacturing engineering will go a long way to maintain the new concepts developed.

The product structure shown on Figure 122 portrays the answer to one company's problem in structuring rotary motion equipment. Prior to preparing this family tree structure almost all the parts shown here were issued at one time. To the oldtimers, who had memorized the assembly and didn't look at the drawings anyway, it did not seem important. You can imagine the surprised newcomers, however, when such a vast quantity of parts were sent to the assembly floor. It required reidentification of all parts prior to assembly. The new product structure is now used to issue parts in a reasonable, orderly, timely, and economical manner without the need for reidentification.

To review, the need for a product structure also requires absolute coordination between a general product structure, a specific product structure, an assembly drawing, and the corresponding detached parts list as shown in Figure 123.

Thus, the term "product structure" covers a significant and important subsystem to an engineering and manufacturing system.

Figure 124 shows a duplicate of the first general product structure.

A specific product structure is also pictured. Notice that not only is the structure shown with part numbers, but an assembly drawing with a detached parts list is shown. Note that the specific product structure does not make an assembly drawing mandatory. Many instances occur where a parts list only is sufficient.

We have previously mentioned the documentation required to make the product structure and corresponding bill of material an effective set of documents. In addition, the drawings and their content will assist the entire system significantly. Not only is an assembly drawing shown, but individual drawings for each part in the assembly are pictured. In previous chapters, we discussed the drawings, drawing content, and

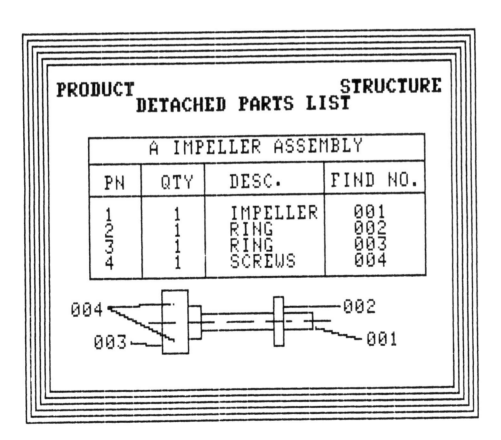

Figure 119 Use of a parts list detached from the assembly drawing is mandatory for accuracy of data.

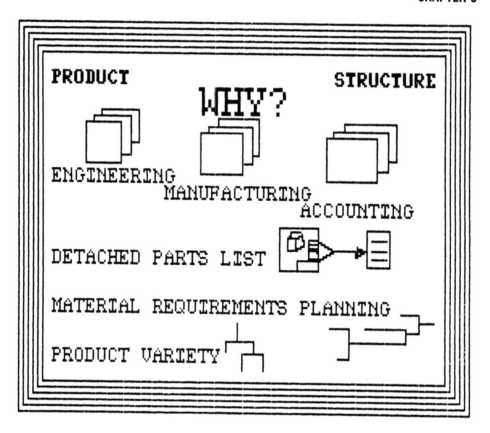

Figure 120 Several important economic advantages are benefits to utilizing a coordinated product structure.

need for individual piece part drawings. Some important and succinct rules for good product structures are listed in Figure 125.

1. The product structure must equal the drawing which must equal the parts list.
2. All drawings must have a part number, but all part numbers do not imply or require a drawing.
3. Product structures must represent the "as built" condition.
4. Product structures for similar products must be identical.
5. Due to modular design concepts, the sales order documentation may need to be considered the top level of the product structure and bill of material.
6. When modules are designated, they must be identical to a sales order input, master schedule items, and design engineering documentation as well as being accepted manufacturing entities.

For just a moment let us look at all the potential data formats available in a mechanized bill of material system. Figure 126 shows these formats divided into four categories:

(1) bill of material, (2) used on, or where used, (3) data file extraction, and (4) bill of material explosions. Process information, engineering change data, standard units of measure, manufacturing operations, spare parts, and effectivity information are but a few of the resulting kinds of data available. Many people believe the most significant system within an integrated manufacturing system is the engineering bill of material. Certainly, if it is inadequate, downstream systems suffer.

With most of the basic rules mentioned for product structures, it is now necessary to discuss a technique which has been gaining in significance and importance for the last 10 years or so—product variations.

Inevitably, a company with a successful product finds it varies continually, either in size, material, or components. These variations may cause considerable problems in documentation. It is possible to find a comparatively simple product with 200-300 variations. To understand the problem Figure 127 introduces the first of some solutions to this problem.

Let us examine a typical product structure and point out some examples of places in the product which consistently vary. As can be seen in this picture, the variations can be simple parts or an entire assembly. We will use this to illustrate the four basic methods by which product variations can be minimized.

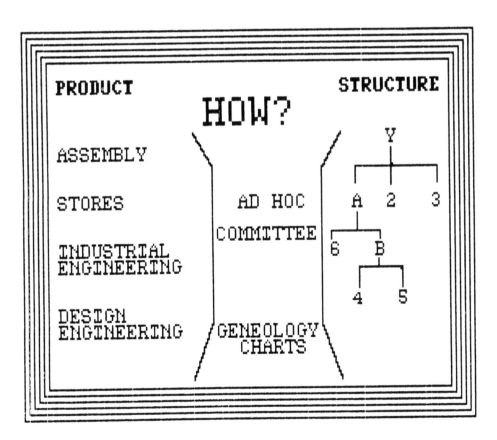

Figure 121 In a company where design and manufacturing cannot agree on the proper structure, an ad hoc committee will go a long way to resolve the problem.

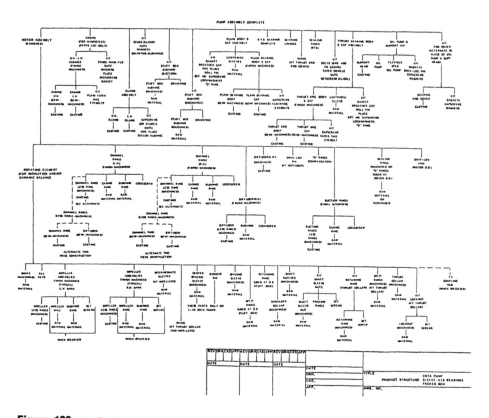

Figure 122 A final satisfactory product structure will often be quite complex. However, it will always be more simple than the previous undisciplined structure.

Attachment Method

The first method for resolving variables in a bill of material is to extract them from the product bill of material, and make them "attachments," as shown in Figure 128. In this example, the variable parts are shown in a separate parts list with a dotted line joining it to the product. Obviously this simplification might actually become a series of assemblies and "kits" of parts.

The dotted line is here used to show that the attachment is neither a physical nor documented part of the product bill of material. Rather, it requires two part numbers to generate a complete bill of material for a complete product.

The automotive industry uses this method. Air conditioning, special radios, special tires, and other "extras" are all considered "attachments." These add-ons are always a positive bill of material in that the addition of the attachment does not require other parts to be deleted.

Increase of Level Method

Figure 129 shows how the variables have once again been separated from the main product and been placed in a separate assembly. The line joining the parts list of the variables to the product structure is a solid line to indicate it is still a portion of the main product bill of material.

This technique is described as increasing the level of the recognized variable. An example of this method is found in documenting wiring lists for computers. The logic wires are in the third or fourth level of a computer bill of material since the total computer, the cabinet, and the wiring board are in levels above them. Moving the entire wire list to the highest level makes it unnecessary to change part numbers of the other levels each time a single wire is changed.

Add-a-Level Method

In some products, an assembly may be composed of parts which always vary combined with parts which are always stable. If the manufacturing operation is enhanced when standard assemblies are known, it may be advisable to separate the variable from the stable parts by adding an additional level between the old assembly and the parts. This is accomplished by adding two new assembly part numbers, as shown in Figure 130. While the amount of documentation of changes is not reduced, recognition of

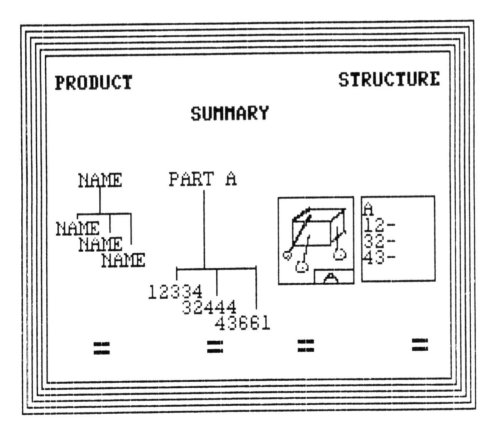

Figure 123 The equality of a general product structure, the specific product structure, the assembly drawing and the detached parts lists makes a bill of material into a successful system. These equalities summarize a significant requirement for utilizing product structures effectively.

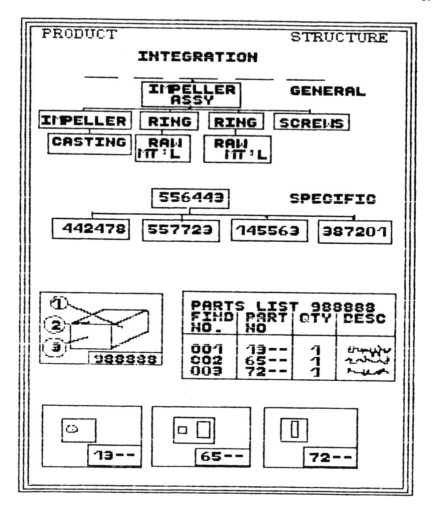

Figure 124 Four structures, the general, the specific, the parts list, and the assembly drawing, combined with appropriate piece part drawings, indicates an integrated bill of material system.

a standard assembly will be the result. An electric motor combined with several sheaves or gear reducing equipment might be examples.

Add-and-Delete Method

Figure 131 is the final example for handling variations. This documentation method presumes a standard product bill of material can be defined. In addition, however, when the bill of material does vary, it is presumed to be only an occasional occurrence. This assumption must become a standard rule, otherwise lack of control will cause real problems.

An example of this method of documentation might occur in the manufacture of typewriters. Suppose the only configuration available has a 110-volt power supply.

A foreign market request for a few of these typewriters with 250-volt power supply would create a bill of material requesting the deletion of the 110-volt power supply and the addition of the 250-volt power supply. For a time this method of documentation might be adequate, however, if the volume of orders for the "exception" increases, it will prove too cumbersome.

With these principles in mind, additions and deletions to the standard bill of material may be made. This procedure allows negative quantities to be generated for parts not required. For special orders, dealing with a job shop requirement from a mass-produced product, the solution is ideal.

We have introduced some new words such as *modules, attachments, variations* and *constants*. Figure 132 diagrams the relationships of these entities. It is intended to show that, indeed, a product can be broken down into these entities which represent logical modules and attachments. No longer is there a series of top-level products with their corresponding part numbers. Rather, a series of similar groups of parts, planned to perform similar functions within a product, but with dissimilar components are represented here.

Once we can picture the various components of a product in a logical way, we are able to begin looking at a powerful tool called *modular design*. The next chapter

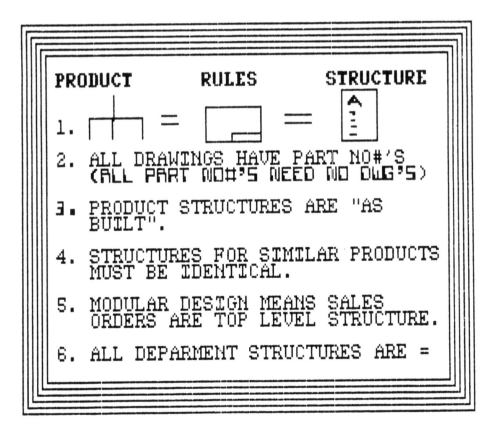

Figure 125 These rules summarize the requirements necessary to structure bills of material effectively.

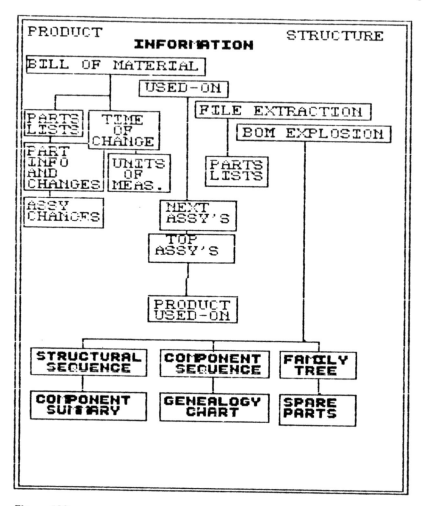

Figure 126   Many different forms of data output can be generated for a bill of material system. Thus, the bill of material becomes the "spine" of all associated systems.

will expand on this subject and show how modular design methods help to minimize product variations, make the engineering bill of material compatible to marketing, and ease the design of new product variations.

Figure 133 demands that we know the economics of modular design. At the top of the figure, sets of potential variations are shown, with each set of these variations tied together with all others into many bills of materials. (For simplicity all the lines for all the bill of material combinations have not been drawn.) If all these combinations were to be documented, the variations would be $5 \times 5 \times 2 \times 6 \times 2$ or 600 product bills of material.

On the other hand, were there to be no product bills of material but, as shown in the lower half of the picture, the variations could be made additive, such as $5 + 5 + 2 + 6 + 2$ or 20 modules documented, the difference would be dramatic. This difference is, in fact, how modular design works. The nonmodular method documents product bills of material as a geometric progression of variations as compared to the arithmetical progression for modular products.

To make modular bills of material effective, of course, there will be no "top-level" part numbers. Instead, it is necessary to select one of each set of module variations and then add as required one each of any attachment needed. The sum of this set of units is then documented on a sales order, utilized as a collection point.

Look again at the new stylized product structure shown in Figure 134, note that the top level no longer says "product," but "sales order." This change is in recognition that the sole consolidation of a complete order requirement takes place on the sales order documentation. This concept is perhaps the best example for emphasizing the need of a product structure and bill of material which can be used by all departments of the company including sales, engineering, manufacturing, and accounting.

In summary, Figure 135, shows the benefits of proper documentation. Increased productivity and reduced expense are a direct bottom-line reward and makes the effort worthwhile. Point-by-point:

1. Manufacturing entities structured to manufacturing requirements.

2. Picking lists for manufacturing leaving neither unwanted stock on the shop floor nor needed parts in the stockroom.

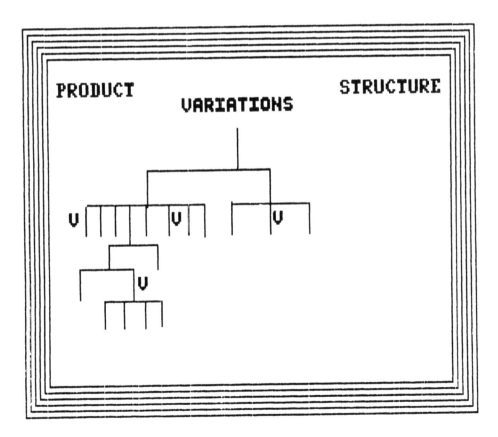

Figure 127 Once bills of material have been structured variations to the product it represents may force another type of change to them to eliminate proliferating them.

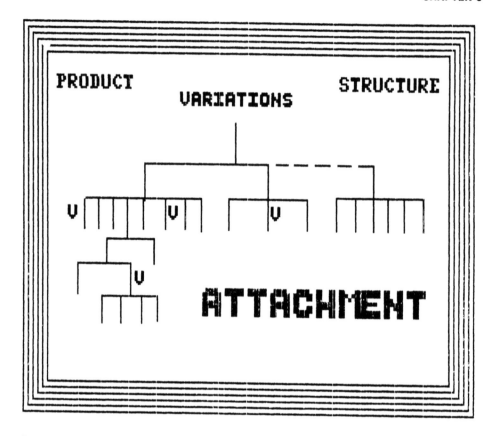

Figure 128 The use of attachment bills of material is one way to eliminate alternate bills of material.

3. Logical assembly sequences are prepared, a welcome improvement to new employees as well as the more experienced.

4. Finished units and their modules are logical entities.

5. Routings for parts and assemblies become more logical, practical, and consistent.

6. Costs for modules and attachments which truly exist may be prepared more accurately.

7. Production control is planning and scheduling real units.

8. Sequential material requirements planning results, such as obtaining the right material at the right time for the right assembly in the right product.

The next chapter will extend the thoughts and principles behind modular design into the technique called *modular design,* an important system and procedure to enhance a bill of material system, and engineering documentation.

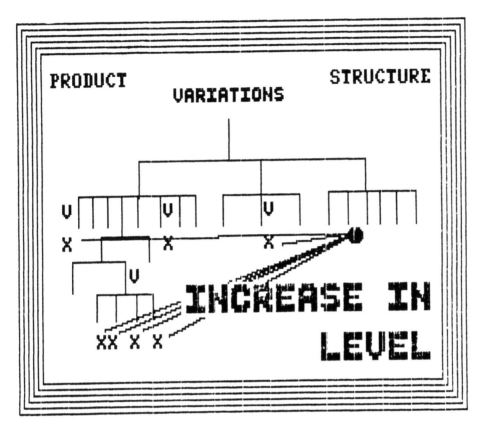

Figure 129 Increasing the level of some portion of the design reduces the number of alternate bills of material required.

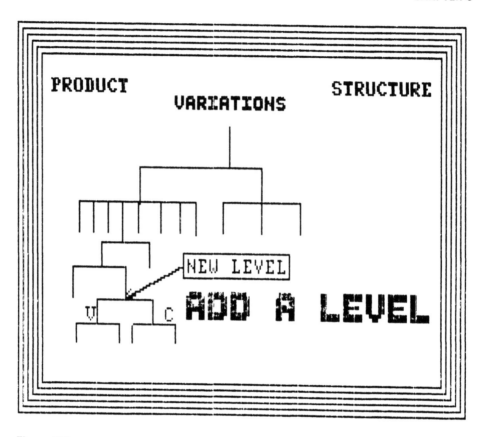

Figure 130 Addition of a level segregates designs that do not change from those which often do within a bill of material.

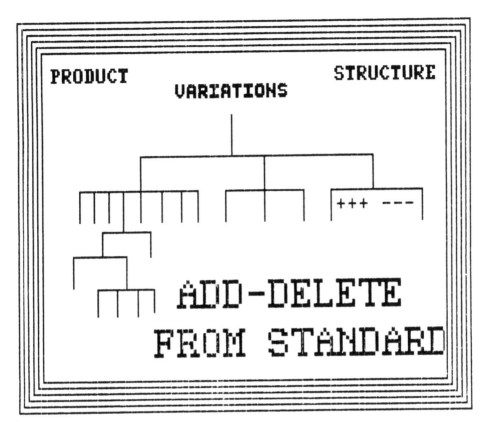

Figure 131 Additions and deletions to a standard bill of material are a rapid, but danger-ous, method to alter a bill of material for a specific customer.

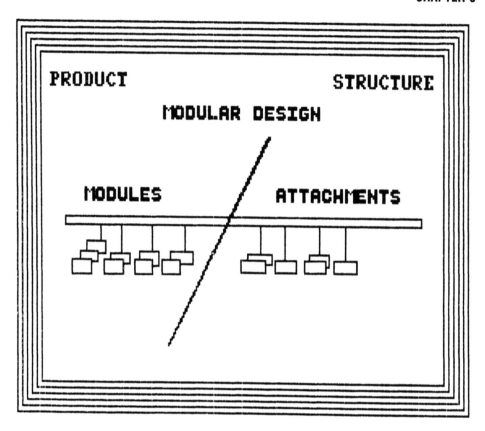

Figure 132 The ultimate documentation of variables is obtained by utilizing modular design techniques. This diagram represents the segregations of modules and attachments.

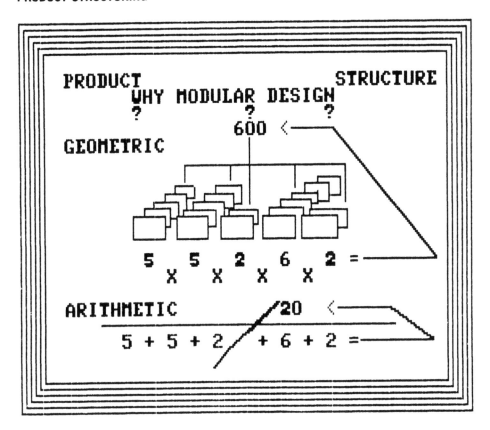

Figure 133 Modular design documentation creates bills of material arithmetically. Variations to designs otherwise creates bills of material geometrically.

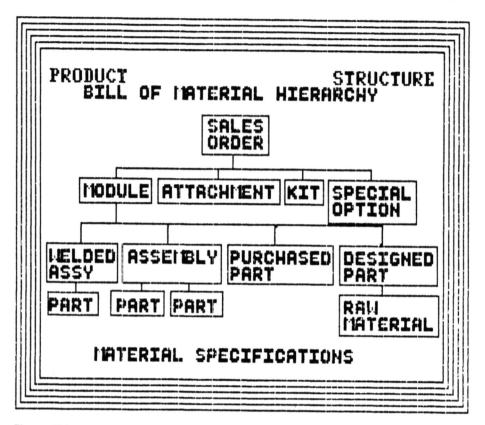

Figure 134 A general modular bill of material is pictured. One obvious change is that the top level is now an order, not a part number.

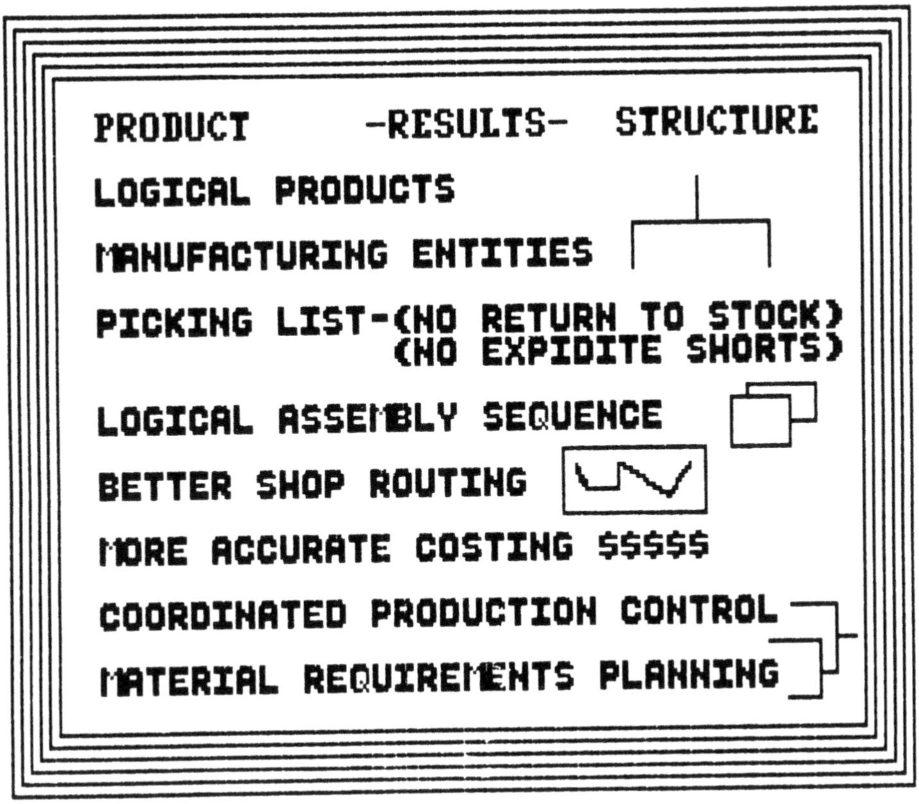

PRODUCT   -RESULTS-   STRUCTURE

LOGICAL PRODUCTS

MANUFACTURING ENTITIES

PICKING LIST-(NO RETURN TO STOCK)
              (NO EXPIDITE SHORTS)

LOGICAL ASSEMBLY SEQUENCE

BETTER SHOP ROUTING

MORE ACCURATE COSTING $$$$$

COORDINATED PRODUCTION CONTROL

MATERIAL REQUIREMENTS PLANNING

Figure 135 The results of effective product structures is this important list of benefits.

MODULAR DESIGN

9

Modular design principles are the natural extension of product structuring. As we develop these principles, keep the picture shown in Figure 136 in mind. It represents the constant conflict between different sets of rules in the procedures of structuring.

For example, a product might be represented with a list of piece parts, showing no structural relationship between any of them. Very simple products may not require more. A product may also be represented at the other extreme where a complete bill of material for each variation of the product is prepared.

There is another solution which develops the method for finding a level in a product bill of material that represents common usage of groups of parts. These groups usually are variously named "modules," "attachments," "options," and "special options." Several may seem unusual until the entire principle is in perspective.

It is appropriate to review some of the basic rules introduced in Chapter 8.

In the preceding chapter four methods were shown by which variables are segregated in a bill of material to minimize changes to them when customer requirements change. Parts and/or assemblies can be isolated as attachments, they can be placed in the bill of material at a higher level, an additional level can be inserted to segregate "variables" and "constants" from each other, and a standard bill of material with additions and deletions can be prepared.

As we will see, each of these alternate solutions require some rules and disciplines not only in bills of material, but also in related disciplines such as order processing, material requirements planning, and drawing content.

In addition to the four ways for separating constant parts and assemblies from those which vary, the product structuring bills of material discussed a comparison between the ancient pyramids and a product structure.

Two definitions were presented:

Product structure A graphical representation showing the hierarchical relationships and combinations of parts and assemblies in a product.
Bill of material A complete record of all parts and assemblies required to make a specific product. It is the paper counterpart of a physical entity.

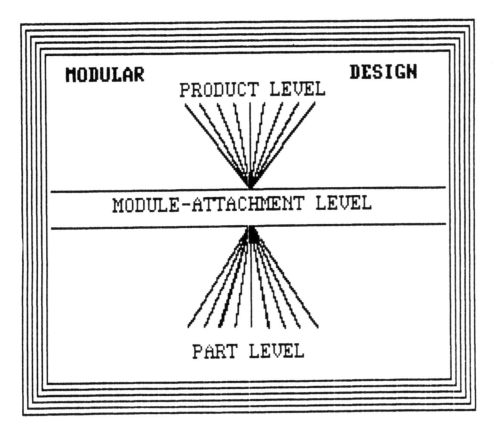

Figure 136 The illustration pictures the conflict between documenting without a bill of material and thus creating long lists of dissociated parts, or documenting each design variation with a new bill of material and creating long lists of bills of materials.

We also pictured modular variations listed on one side of a diagonal line and attachments on the other. We will enlarge upon this concept to begin developing the methods and needs for modular design.

As more companies systemize and convert to such systems as material requirements planning, shop scheduling, and production control, the imperative of an accurate and timely bill of material is emphasized repeatedly. Figure 137 shows different representations of bills of material known by such names as stock lists, pick lists, parts lists, pull sheets, shop order pages, and material lists.

Most all had errors and were hardly ever current. It was only when they were converted to files for computer systems that the significance of these inaccuracies became apparent.

Some of the many problems found with the present mechanized bills of material are:

Product variations

Product proliferation

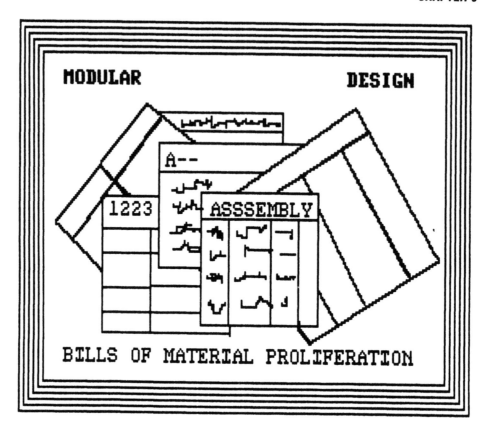

Figure 137 Proliferation of bill of material methods, formats, and requirements is a natural phenomena both within and between companies.

"Phony" assemblies

Multidepartmental bills of material

Nondetached parts lists

Tabulated assemblies

Continuous "restructuring" for marketing

Inadequate records for master scheduling

Often an uneasy feeling arises that "something is wrong." The rules we will cover here we hope will start anyone with such feelings on a path to make their bills of material respectable, accurate, and a powerful management tool.

Early mass production products such as firearms, farm equipment, and automobiles, used a standard bill of material. Variations to the product were almost nonexistent. As Henry Ford said, "you can have any color car you like as long as it's black."

The bill of material or stock list was a part of the final assembly drawing, and, because the first products were assembled on a continuous assembly or conveyor line, few or no subassemblies were defined or required.

Soon, however, most successful companies found their customers would buy additional products from them if different styling, larger sizes, greater capacities, and other product diversifications were only available. These variants were successfully marketed, and thus, more and more bills of material were prepared. Each new design proliferation made it increasingly difficult to find old designs from which to vary, and thus, variations to variations caused a serious increase in the number of drawings required. The proliferation schema is shown in Figure 138.

Customers asked for the same basic product but with some variants such as blue, not green, short, not long, with a gear guard, shorter boom, U-dozer not blade, and many others. The list of product variations grew rapidly and increased the popularity and profits of the company, as shown in Figure 139.

Companies such as Sulzer Brothers in Switzerland with thousands of weaving machine variations, or Caterpillar Tractor Company in the United States, with 30 major variations of their D-8 tractor alone, took pride in the popularity and flexibility

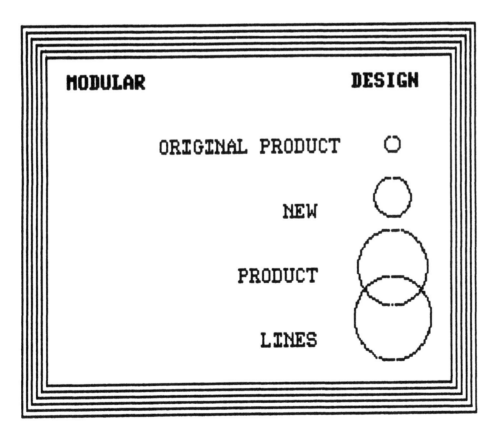

Figure 138 · Design proliferation began if a product was successful. The next customer wanted it bigger and/or different.

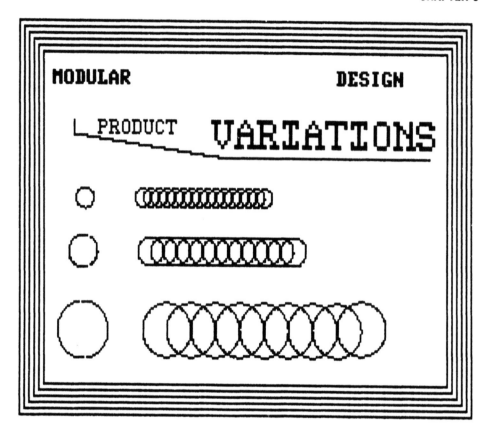

Figure 139 The result of customer requirements was the generation of multiple product lines each varying to meet a special requirement.

of their products. Other companies were not so fortunate. Some companies went out of business, unable to cope with the many variables. Others found themselves relegated the task of "customizing" in a job shop with "specials" no one else built.

Manufacturers began to see the need for designing standard assemblies which could be put together for a customer order, thereby creating "standard customizing." This only caused the problem to appear at lower levels in the production process since more assemblies were documented to place the parts into logical groups. As more levels were required and more groups of parts designated at each level, an indented or structural bill of material was required to keep track of these units.

Tabulated assemblies were often used to picture variations of assembled parts. However, during the period beginning about 1938 through 1958, data processing using unit record-punched cards became popular. Bulky and slow small-scale computers also surfaced during the 1950s to alleviate the documentation problems. Unfortunately, this emphasis on utilization of these methods was aimed at ordering parts and assemblies more quickly, not at easing bill of material documentation problems.

The first mechanized bills of material were prepared on either punched cards or paper tape. In many cases, this project was activated unilaterally by the data processing department as both engineering and manufacturing looked on sceptically. Concurrently accounting was mechanizing bills of material for costing purposes.

Data processing departments attempted to sell mechanized bills of material to manufacturing by showing them how rapidly a computer could summarize total part requirements for each product and then add them together for all products for both purchasing and manufacturing requirements. At first, these efforts could only generate gross requirements. Thus, the structure of the bill of material could be mostly ignored. In fact, a good summarized bill of material for each product was the only real requirement.

It wasn't long before requirement planning techniques were introduced. This created a new set of problems. Work-in-process and inventory had to be mechanized so that net requirements could be computed. In addition, the requirements had to be time-slotted on a weekly or monthly basis. The structure of the bill of material took on new importance . . . and so did engineering design changes.

Most companies soon found that their engineering department was not going to be a partner in the mechanized bill of material process. However, once manufacturing had their appetite whetted, they became the prime mover in mechanizing bills of material.

Some companies found themselves relying almost completely on mechanized bills of material to order raw materials, parts, and assemblies. They became aware of the extreme complexity in continually interpreting the engineering department's handwritten documentation and converting it into useable, mechanized information. Requests to engineering increased. They were asked for improved documentation as well as greater data accuracy. Together with more sophisticated ordering procedures such as *net change,* these new requirements have placed an excessive strain on engineering documentation. Implementation of modular bills of material will, we believe, alleviate this problem logically, practically, and economically.

Variations, sizes, and extras created multiple bills of material. Figure 140 shows five different bills of material to picture variations to the product. As mentioned previously, hundreds of variations to a particular basic product might exist. If, as these variations were documented, the drawings created were drawing-number oriented not part-number oriented, simple changes to a drawing could change literally hundreds of part numbers for parts identical to previously designed parts.

This unfortunately created the false impression that a majority of parts in each new order were new. The next step in this misconception was to create a system of documentation presuming that all designs for the next order would be new. This principle, in turn, created the need for a manufacturing system to build one part at a time. This chain of events created unnecessary costs, systems, and an inability to rapidly react to a new customer order.

To change constructively this activity cycle and inbred thinking, a series of steps in a preliminary investigation are required. The investigation is followed by a seven-step procedure for creating modular bills of material and the rules governing bills of material, assembly drawings, and piece part drawings.

The preliminary investigation examines and analyzes the present bills of material. In addition, a cursory study of past orders for products (models) may be helpful in order to identify the quantity of variations which have been documented, as shown in Figure 141.

A product structure is physically prepared. This may seem a costly first step. How-

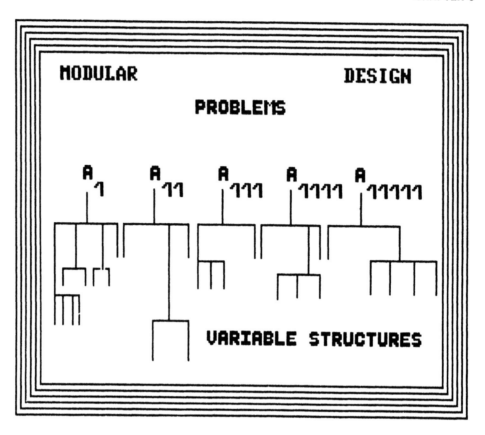

Figure 140 The result of design proliferation was seen in the paper files of hosts of bills of material. Mechanization of this data created excessive storage requirements, and the data was difficult to find.

ever, it is mandatory to establish what presently exists. No matter how many years experience the examiner has had with the product under study, preparing the actual product structure holds some surprises.

Three simple product structures in Figure 141 show three different types of product variations. The top structure shows either too few subassemblies, or too many parts and assemblies brought together at final assembly. The structure on the left is meant to represent a product with a biased bill of material due to the size of some of the components, such as in the manufacturing of boilers, aircraft, computers, and cement kilns. The third structure appears to show a well-balanced bill of material for the product.

A cardinal rule is to create a product structure which meets the complete and total manufacturing and assembly requirements.

The product structures prepared should also be used to identify variations in a product and between products.

The next 7 steps will establish the modular bills of material required. They must be followed with a good deal of product knowledge and common sense.

Figure 142 for step one shows some pictures of pump assemblies mounted with electric motors on a base plate. We will use this product as an example. It is first necessary to establish the data source for the bill of materials of the products being studied. Hopefully, the order file itself (whether manual or mechanized) will be useful for this exercise. At any rate, each bill of material must be examined and compared to all others to identify the variables within each and between each product line. Surprisingly, this examination is not as difficult as it may sound, since designers and drafters are usually skilled enough in this type of examination to expedite this step.

The investigation is performed to create order out of seeming chaos (or standard parts lists from variable parts). This examination must be objective and free of bias. Significant variations such as piping, electrical wiring, or tubing may be set aside at the beginning while the remaining portions of the product are examined.

Once the variations for a series of products have been identified, the second step is to subtract these variables both within and across product lines, as shown on Figure 143. For products which are extremely complex it may be necessary to restrict the study to one product line. Remember this as the rules are presented.

Variations can look extremely comical depending upon the designer who created them. A simple bracket may have many variations each completely adequate, but

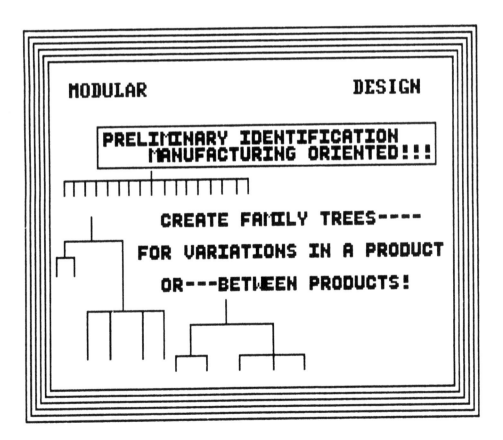

Figure 141 To convert bills of material to a modular format some important preliminary steps are required.

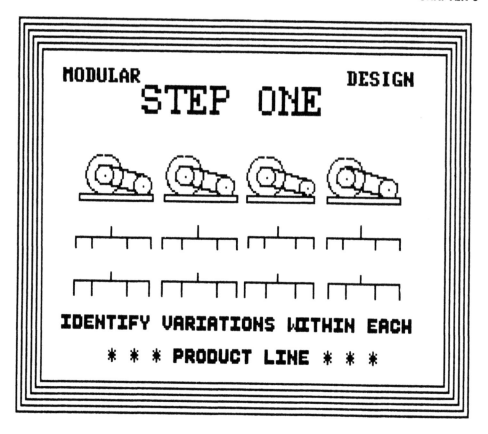

Figure 142 Step one separates constants from variables.

requiring different hole spacings, or some other change. A battery may vary as one part because it may be purchased wet or dry, 12 or 24 volt, foreign or domestic, etc. Some variations must be set aside to determine whether a more standard design could replace them all and thus eliminate the variation completely.

Figure 144 shows the separation of "constants" from "variables," major and minor logical groups of parts which repeat from order to order can now be identified. A brief set of definitions may be helpful at this point;

Module   A *major* assembly a variation of which is always required in a finished product.
Attachment   This may be used to modify a module of a basic product. An attachment may modify the product as 110 volt versus 220. An attachment may alter or be added to another attachment.
Option   A variation which has special or irregular usage.

More detailed definitions are required once modular design is underway. Each company's products dictate the most appropriate. Remember that the master schedule, product forecasting, order input, and engineering documentation will be based on the segments of products created, so the definitions used should be short, clear, and concise.

With modules, attachments, and options defined, step four requires a logical sequence of selecting variations to make a finished product. For example, a motor and engine base selection such as A or B in Figure 145 allows the selection of one of two pumps X, or Y. In turn, parts L, M, N, and/or K, P, R, are needed for the final product.

This selection, however, may be incorrect and illogical if in the pump business. The customer may select pump capacity as the major criteria. Thus, pump selection must be first, followed by an orderly selection set such as horsepower, voltage, speed, environmental conditions, etc. The decision table will be built in a manner so that both sales and marketing can fully utilize the decision-making process. In some cases the selection of a pump or motor disqualifies certain additional variations. The decision table must be clear in this type of judgment. It should be emphasized that modular design techniques make these decisions clearer to see, and, in fact, may clear up any confusion existing in order processing.

Step five suggests altering the physical product. Punching or drilling additional holes, creating symmetrical parts, relocating lugs and knobs are all examples of physical changes, as shown in Figure 146.

More complex changes are often necessary. For example, electrical wiring may permeate a product, but require different routing due to product changes or different

Figure 143 Step two essentially segregates the different constants and variables between product lines.

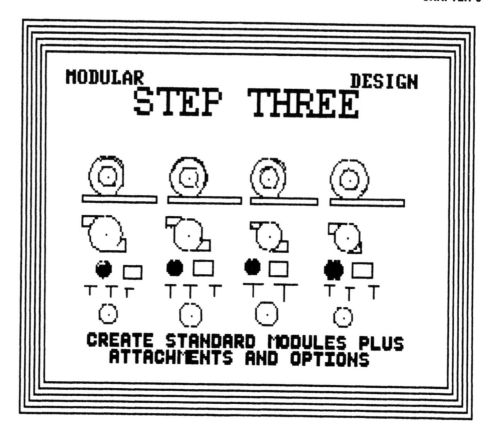

Figure 144 Step three identifies each portion of a design as being a module, attachment, or option.

wiring due to electrical characteristics. Modular wiring harnesses may be needed. Increased capacities in iron ore pelletizing dryers may require the design of a modular sectionalized conveyor system. Roll conveyors may require several sets of cross-braces and extra holes in the conveyor side frames.

It is important to remember that the physical product and the modules defined must be logically equal. No greater disappointment could be found than assuming the preparation of paper modules and attachments will alleviate or bypass the inadequate design principles. In other words, a paper solution only is probable not possible.

With modules, attachments, and options defined and a sequence of decisions prepared to combine them into logical finished products, the next step, six, should be relatively easy, as indicated in Figure 147.

The sales order form whether manual and/or mechanized will become the "collection point" of all the specific variations required to make the specific product for the specific order. Perhaps the best way to express this collection point is to define the order number as the top-level part number of the product variation which is required.

Not only will part numbers for modules, attachments, and options appear on the order, but special instructions, tests, specifications, and other order-oriented data appear. It is also possible to consider this document to be the one which compiles standard data with nonstandard data as in a job shop manufacturing atmosphere.

Step seven, as shown in Figure 148, may seem to be unimportant compared to the others. However, many companies have avoided or exaggerated the product variation problem by creating one standard bill of material and hundreds of "add" and "delete" documents. Compounding the problem is the practice of layering adds and deletes to the adds and deletes. A veritable maze of paper-work is created. Eliminating this practice is mandatory.

Once the rules and the spirit of the rules have been established, limited use of additions and deletions to a standard bill of material will be a rapid and economical tool to issue one-of-kind variations quickly. Eternal vigilance is the price of such liberty.

In the previous chapter on product structuring, Figure 134 pictured the type of bill of material resulting from modular design. A second look at this figure will be helpful now.

The rules required in other portions of the engineering documentation procedure become extremely important to ensure an effective system. The first set of rules we will introduce relate to the bill of material, as shown in Figure 149. These rules should

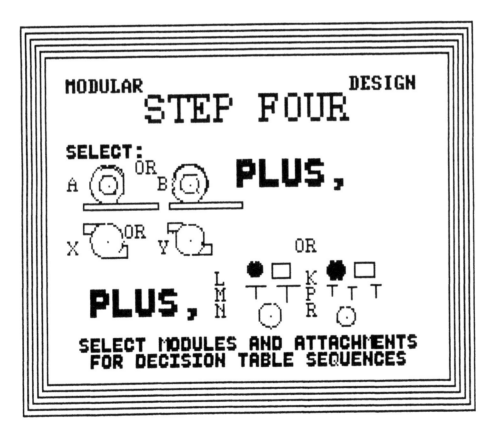

Figure 145 Step four reformats the product into logical entities for selecting the proper configuration for a customer requirement.

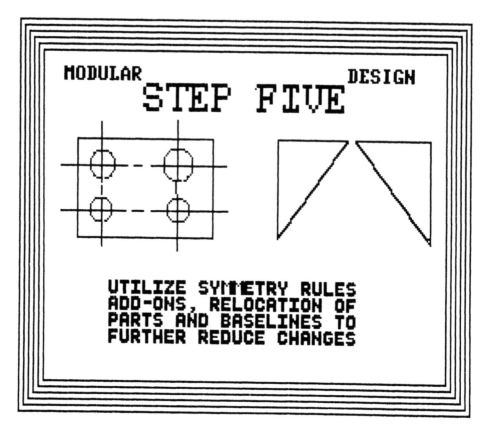

Figure 146 Step five indicates the need to alter the physical design of the product as well as the documentation.

by no means be considered all-inclusive. The specific rules assume a reasonably logical and accurate bill of material system is in existence.

When modules and attachments or options are combined for a specific customer order, the combination must be positive. In other words, adding one to another is implicit—at all times. If there is a condition of "same as but," or "subtract part A when adding this feature," it will soon be impossible to know what vagaries have been imbedded into the bills of material, and confidence in them will deteriorate.

On the other hand, no attachment should be so obscure that it is impossible to determine where to put it. For example, if a pump impeller, a crankshaft, an electric motor, or a dozer blade were the attachment no problem would be apparent (unless, for example, 100 different electric motors are all attachments). However, a bracket, a hydraulic reservoir, or a wiring harness may cause difficulty.

As we established a procedure for adding an attachment to a module or other attachment in a positive way, so must we add one standard module or bill of material to another without changing either of them, as shown in Figure 150. We must also recognize that an attachment may have attachments due to the hierarchical nature of products. For example, a crane may have an enclosed cab as an attachment. In turn, the cab may or may not have electric or hydraulic windshield washers. The washers may or may not have a window-washing unit with them. Each product, however, has its own set of possibilities, so care must be taken to assure resolution of them.

We cannot emphasize enough the accuracy required of the actual product structure as defined by an assembly parts list. The assembly parts list should have previously been prepared based on the product structuring procedure. Should accuracy deteriorate, the module design concepts presented here will erode integrity and credibility.

The next set of rules are concerned with assembly drawings, as shown in Figure 151. Many of these rule violations may have been resolved during the investigation of product structuring, but are worth reinforcement.

Tabulated assemblies are the antithesis of modular design. Since a table of variables must be tied to the drawing number upon which they appear, this requires a dash number system for the drawing at least, and makes the addition of new modules or changes to slightly different modules difficult and/or expensive.

A bill of material's sole reason for being is the "structure" to show that parts and assemblies appear not only in the sequence of actual assembly, but also in the sequence of manufacturing or purchasing. Thus, an attempt to prepare a "common" parts list is time consuming and unnecessary. The material requirements planning system is the proper method for summarizing all parts, and not for just one bill of material, but for all.

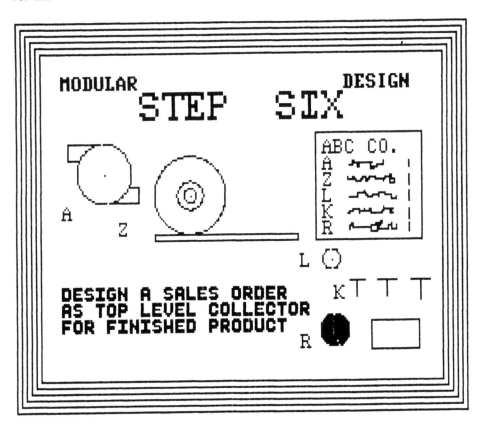

Figure 147 Step six recognizes the sales order as the collection document for a final product configuration.

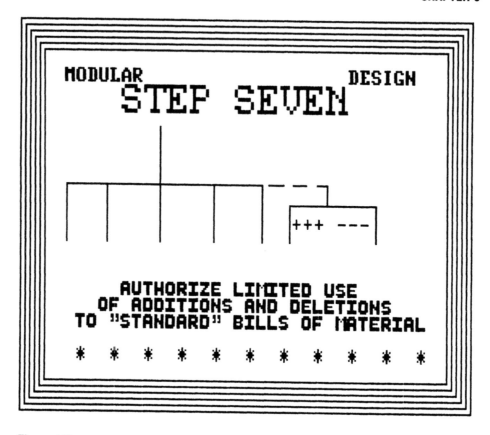

Figure 148 Step seven cautions against the use of additions and deletions to standard bills of material.

Since the assembly parts list is now a part of a mechanized bill of material system, it is not only unnecessary to have a parts list on the drawing, it is actually quite dangerous. If a parts list appears on the drawing and is also in the master file of the bill of material system, two "masters" exist.

Figure 152 shows additional rules for assembly drawings. Since attachments can be designed for a variety of reasons it is important to define them if possible. Those attachments rarely used become options. Other designations such as stand-alone or standard, may be feasible designators. The master schedule requirements should also be considered when defining these names.

Serviceability may seem a strange subject to discuss with regard to modular design. However, it seems reasonable to believe that not every part on every level of every bill of material should be serviceable from an economical viewpoint. For example, the bolts and nuts of a simple assembly may be too expensive to show as spare parts, but the entire assembly may not be. Thus, the serviceability level may play an important part in determining just where attachments will be documented and whether they will be treated as spare parts. (Note that many companies make no distinction and show every part on their service parts lists.)

The use of stylized drawings in modular design can decrease drawing quantities dramatically. As modules are designated, many of them will look identical, and be assembled in an identical manner. No assembly drawing need then be made for each of these assemblies. Rather a "stylized" drawing may be prepared and then listed on the parts list for the assembly. The part number of the stylized drawing is always different from the assemblies for which it is drawn, and the stylized drawing has no parts list of its own. One company reduced its assembly drawing output by 50 percent using this technique.

The last set of rules are related to piece part drawings, as shown in Figure 153. Incidentally, it is important to consider possible major changes in creating piece part drawings. However, the rule, "each drawing must have a part number," does not have the corollary that each part number must have a drawing. In past chapters the entire subject of piece parts and part numbers were discussed.

Since documentation within a company should not be subject to changes by the whim of some external variable, the company documenting parts should do so with their own independent numbering scheme.

Even though each purchased part must have an individual part number, when it has some change made to it after it is received, another new part number must be assigned after such alteration. Thus, there will be one number for the nonproprietary

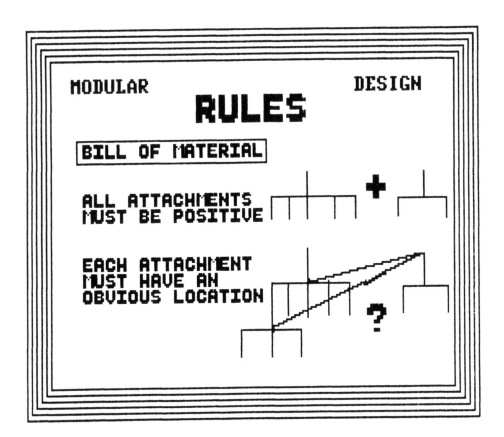

Figure 149 These documentation rules for bills of material are required to implement modular design successfully.

Figure 150 These documentation rules for bills of material are required to implement modular design successfully.

part as purchased and another for the new proprietary part after internal alteration. While these rules appear to create additional paper work (if they haven't been followed previously), an investigation into present methods for keeping track of parts before and after alteration which have the same part number will show a substantial cost for present documentation.

Since parts will be documented for various reasons, it is important to be able to find them in their various categories. A part type code has been an effective means for this segregation, as shown in Figure 154.

Parts are not only individual pieces, they are weldments, fabricated assemblies, modules, attachments etc. Some of this terminology seems overlapping, and will be unless each term is specifically defined. When kits, reference drawings, and stylized drawings are used, they too become separate categories of parts drawings.

Since a rough forging or casting may be used for many different finished parts, a separate part number should be assigned to each rough forging and casting. In addition, the purchased raw materials should each be given a part number, since many different piece parts can be made from the same raw material. Material requirements planning systems will find this data useful to combine part requirements and summarize the rough casting, forging, or raw material requirements.

Figure 155 shows rules for piece part drawings. While piece part drawing should be prepared for each part, use of a tabulated drawing to combine them is emphatically not recommended. We have previously covered the problems of this type of drawing. However, as mentioned before, a classification system performs a more efficient way to accomplish what a tabulated drawing was supposed to perform, and these modern retrieval systems perform rapidly and effectively. Computer-aided design/computer-aided manufacturing (CAD/CAM) systems also are inhibited with tabulated drawings.

For the same reasons, parts should never be detailed on the drawing of the assembly for which it is required. An overpowering reason for this rule is to be able to use the same part on another assembly if possible. Since the part detailed on an assembly drawing normally carries the drawing number of the assembly plus an additional "dash" number, it is difficult and/or impossible to use such a part on another design.

These rules for bills of material, assembly drawings, and piece part drawings represent some of the most critical guidelines. As mentioned previously, other rules exist. It is our hope this sampling will open the door to a full investigation of documentation requirements if discrepancies exist.

The cost for investigating and developing modular design techniques is easily offset by the benefits. Some follow. Often it is found that many bills of material need major

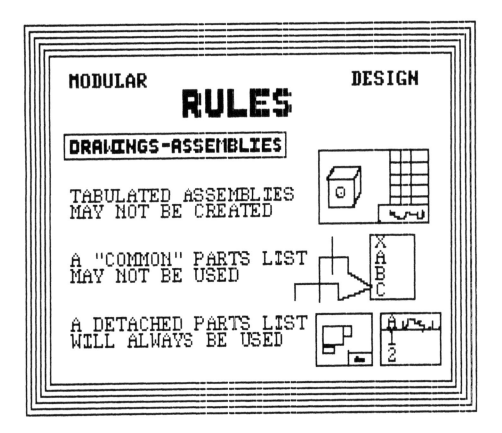

Figure 151 These documentation rules for assembly drawings are required to implement modular design successfully.

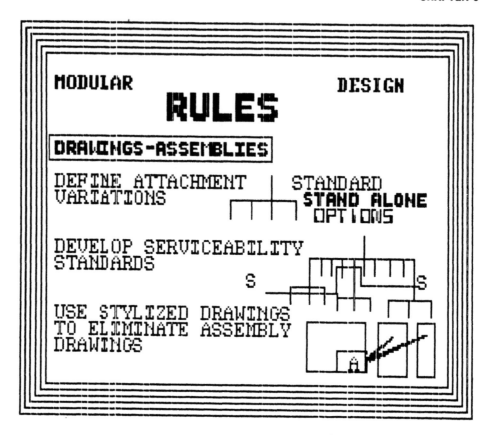

Figure 152 These documentation rules for assembly drawings are required to implement modular design successfully.

revision. Fortunately, no products have been made with totally poor product structures, so modular design usually can be compared to pruning a tree, not cutting it down. Also, some of the rules, such as the use of stylized drawings to represent many assemblies, will reduce drafting and design time if used judiciously. However, some significant reasons for modular design appear when examining an integrated system, as shown in Figure 156.

The sales order input will be more rapid and accurate if it is coordinated with engineering bills of material. Engineering changes will be more rapidly and accurately reflected in the price book.

The master schedule accuracy depends upon the logical entities scheduled. These units will be more accurately portrayed if they are the units marketing and sales use daily.

Forecasting is a difficult and serious business—and expensive. It will be helpful to forecast in units that are meaningful to everyone involved.

If a company builds to order or to inventory, delivery schedules will be enhanced if some parts of a product are maintained in inventory. These parts will be chosen and maintained more precisely to requirements if they are modules or attachments known and understood by all personnel concerned.

Design entities which can be used by these people means engineering does not spend so much time interpreting every order received. This inevitably increases engineering productivity.

Manufacturing has for too long been forced to restructure bills of material in order to manufacture the product. They have also then been forced to interpret each engineering change, make exceptions on pick lists, and generally redesign the documentation. This has been most costly.

Last but not least, the cost of a product will match pricing for the product sold, designed, and built. Since productivity for profit is a major goal of modular design, we could not leave the subject with a better reason for subscribing to the procedure.

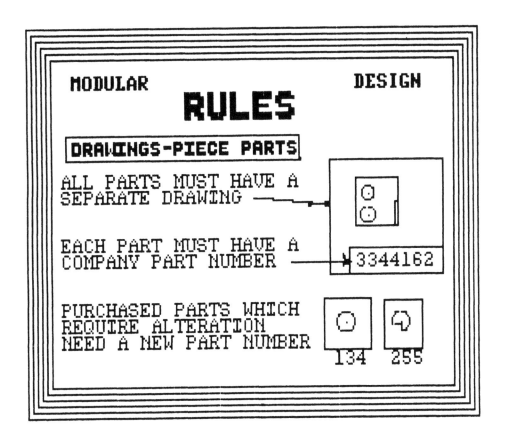

Figure 153 These documentation rules for piece part drawings are required to implement modular design successfully.

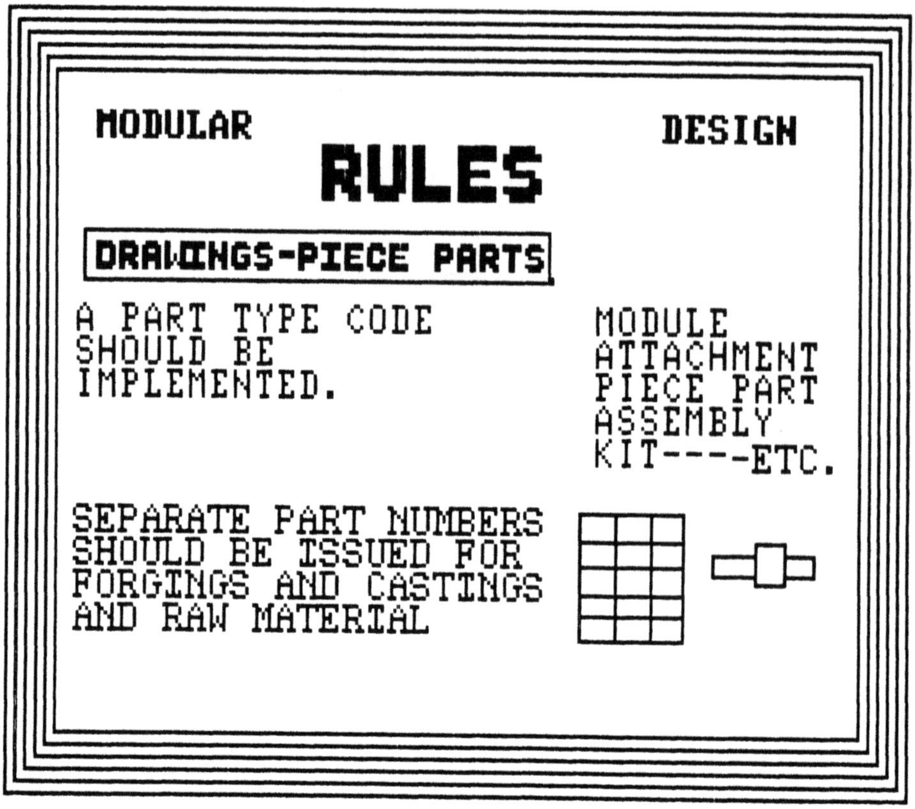

Figure 154   These documentation rules for piece part drawings are required to implement modular design successfully.

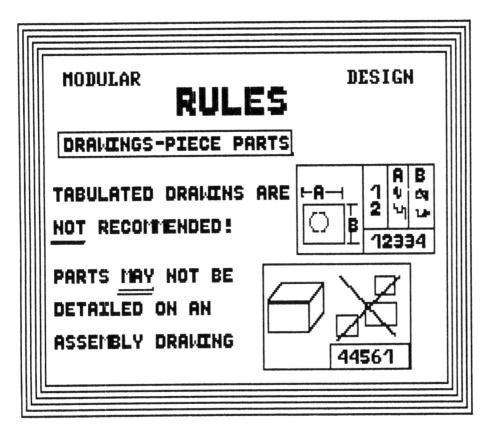

Figure 155 These documentation rules for piece part drawings are required to implement modular design successfully.

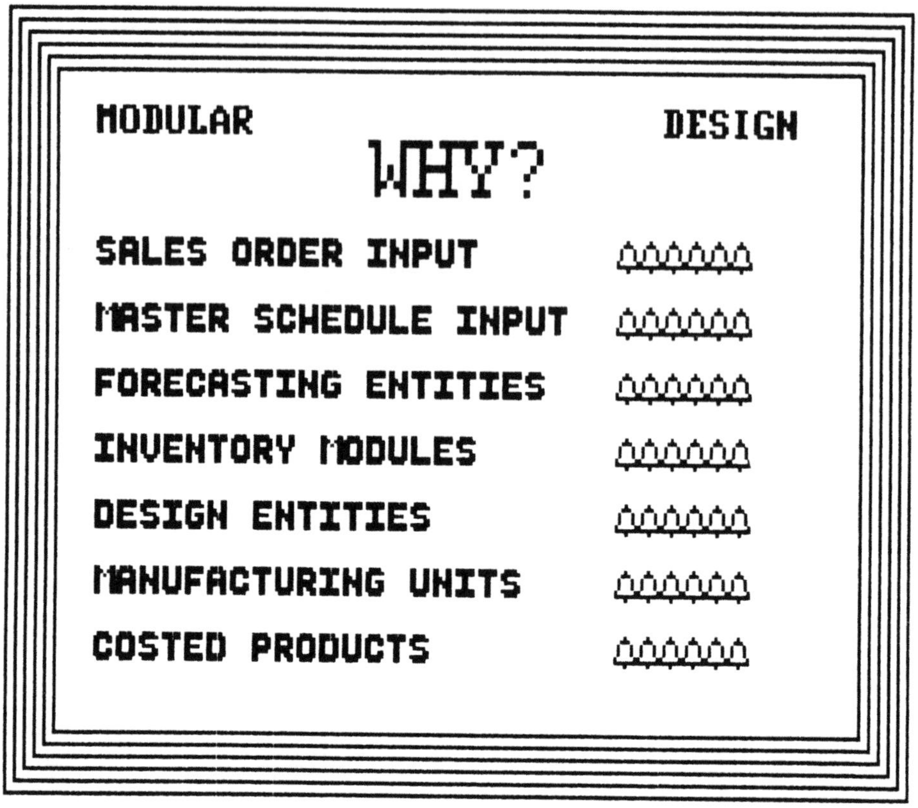

MODULAR   WHY?   DESIGN

SALES ORDER INPUT

MASTER SCHEDULE INPUT

FORECASTING ENTITIES

INVENTORY MODULES

DESIGN ENTITIES

MANUFACTURING UNITS

COSTED PRODUCTS

Figure 156 The economic benefits of implementing modular design techniques are listed.

JOB SHOP VERSUS MASS PRODUCTION

10

JOB NUMBERS VERSUS PART NUMBERS

In the past several chapters we have made no distinguishing remarks between documentation of companies who are admitted mass producers, and job shops. The distinguishing differences between them have become hazier over the years since customers have become more demanding of changes even to standard products. Generally, however, it appears to be that some specific manufacturers can be classified as job shop one-of-a-kind producers. Some examples of this type of manufacturing are:

Fossil-fired boilers

Nuclear reactors

Special steel mill products

Large gas turbines

Large compressors

Job shops often have a completely different viewpoint on what constitutes a documentation system. Almost all systems are customer and job-number oriented. Many of the part number systems we discussed came from such a company. If job numbers are "king," then engineering documentation takes a back seat. Job number manufacturing means an identical part may be found, duplicated, and manufactured many times under different job numbers. Unfortunately, this identification system over the years "proves" that duplicate parts and use of identical parts does not occur.

With the advent of computers, the problems of a job order system have become more obvious. Bills of material are almost useless in such a system since the meticulous documentation of engineering is lost under the pseudopart numbering schema found with job numbers. Economical ways to maintain both numbering systems have yet to be developed. In addition, the philosophy of a job number system has created several manufacturing practices which are not particularly economical. Engineers within such a company also have some misconceptions which must be disspelled before a change can be made to utilize a modern bill of material system. Look at some of these misconceptions and/or practices which follow: Figure 157 shows six of them. The left side of the figure shows a series of part-number oriented requirements, the right side shows the present practices or results of a job number system.

A.   Converting to a part number-oriented system often creates the impression that drafting costs will increase due to the need for more drawings in the form of piece parts. A careful study will show that, while more drawings are needed, the number of "lines" drawn are almost the same. Even though a single drawing showing all the parts was previously prepared, many detail parts are drawn on the same drawing and

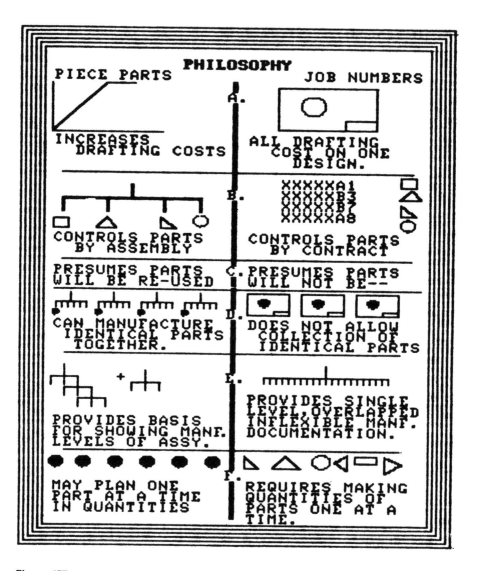

Figure 157 Some of the philosophical differences between manufacturing systems using a job number system versus a part number system are illustrated.

many sectional drawings are also made to clarify the assembly. In addition, as we shall discuss, the piece parts drawn are reuseable.

B.   A piece part-oriented system controls the issue and assembly of parts with a "structure" which takes the form of an assembly parts list. A job number system controls the issue of parts by relating each design to the job number of the customer. This relationship is equivalent to a summarized bill of material. This method achieved accurate costing results in the days when no other means for collecting data for a contract existed. The accounting systems were designed to collect costs in this manner by job number. The bill of material system eliminates this need.

C.   Designing a system where piece parts and part numbers are prepared implies that they are and will be reuseable. A job number system is based on the one-of-a-kind philosophy which assumes each design to be so unique that it will never be used again.

D.   In a part number system, the documentation is prepared so that an identical part, no matter where it is used, can be collected into one batch and manufactured together. The job number system does not allow this to happen easily or often at all since it will require a series of drawings, each documenting several parts, to be collected together and then marking off the parts which will not be manufactured together.

E.   A part number system allows the product to be structured so that manufacturing entities and a level-by-level manufacturing sequence can be documented. The job number system relies on the engineer to document the structure of the product for manufacturing correctly. Quite often, however, assemblies are not documented as such, the result of the job number system is to provide a single level, overlapping, and inflexible list of parts for the product.

Figure 158 is another series of problems and misconceptions of a part number system versus a job number system. Once again, the left side of the drawing shows the requirements or results of a part number system, the right side shows the present practices of a job number system:

A.   Most engineers presume not only increased drafting cost, but also an increase in the number of parts which must be documented. Nothing could be further from the truth. As more and more parts are documented, the similarity of parts increases. With a good classification and design retrieval system, more and more parts become quickly available, actually reducing the number of new parts required.

On the other hand, a job number system requires a continuous stream of new parts designs, even when many of them are duplicates. Multiple parts on one drawing and the system of preparing such drawings means that a change in only one part requires a new assembly drawing. Thus, all duplicate parts are renumbered.

B.   The use of part numbers and the bill of material system allows product costs to be developed at every level of the bill of material. In fact, the use of the bill of material to prepare a *cost roll-up* is a significant savings in the processing of data. On the other hand, job number-oriented drawings may or may not be segmented to obtain such level-by-level costs. Thus, the costing of a product or order is often for large "chunks."

C.   The use of a part number system, particularly if a random and sequential part number system is used requires a design retrieval system. This is really not a disad-

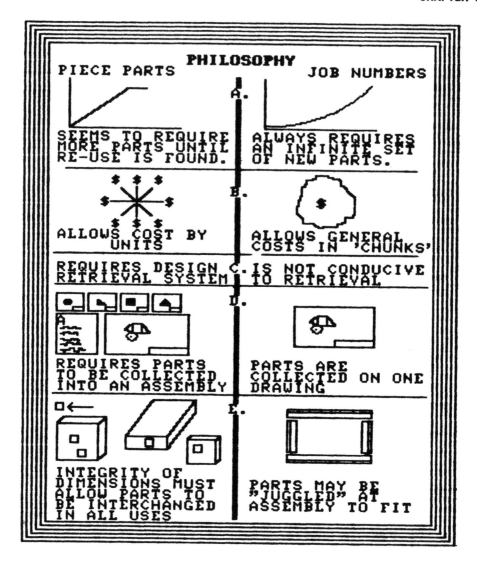

Figure 158 Some of the philosophical differences between manufacturing systems using a job number system versus a part number system are illustrated.

vantage in that retrieval systems create a means for preparing families of parts and corresponding data base for CAD/CAM. A job number system and multiple drawings are simply not conducive for a design retrieval system.

D.   Once piece parts are prepared on individual drawings, they must be collected to document them as an assembly. Thus, an assembly parts list must be prepared. The bill of material is used to prepare this document. An assembly drawing is also necessary to show how the parts are put together. The job number system documentation groups the parts on one drawing.

E.   Since piece part drawings have been prepared so that they may be used in many different designs, the dimensions, tolerances, and other criteria for each part must be of an integrity such that it will be manufactured to fit any application. On the other hand, a multiple part drawing allows some give and take in the manufacture of parts as they may be "juggled" to fit at assembly, and the fabricator can see where mistakes may be made or be less accurate in the parts made.

All of these examples have been used by personnel in job shops either to prove that a job number system should not be replaced, or to prove a part number system is too costly. Company after company, however, has studied these "facts" and misconceptions, and those who have are converting to part number systems. Thus, it can be said that many companies are converting from a job number system to a part number system. No company in our experience has gone the opposite route.

MATERIAL PLANNING IN A JOB SHOP

If we can establish that a part number system and a bill of material system will work effectively in a job shop environment, the next question which must be addressed is the part that material requirements planning will play in such a company. Material requirements planning (MRP) is thought to be useful in only those companies with a large volume of identical products and with companies who can also forecast their product requirements accurately. Since a job shop usually has neither of these luxuries, can the techniques embodied in these systems be used?

Fortunately they can if we consider orders as the forecast, and an "order" bill of material as the product bill of material.

Even those companies who mass produce are finding problems with MRP, because they too find their customers tend to require customized products. Such methods as modular bills of materials, product structuring, and others are used to manipulate the data which customized requirements create. The disciplines of accurate inventory, mechanized bills of material, purchasing records, work-in-process, routing systems, and others simply are not satisfactory enough to make MRP work well. These disciplines are notoriously poor in a job shop since parts tend to be ordered for one customer order, part numbering systems are inadequate, multiparts on one drawing make it difficult to reuse parts even when they are found. Expediting is often done with accounting controlled job numbers for batches of one. Each and every weakness or lack cited above are primary requirements for MRP. Shall we abandon all hope and find some other way? It is the hope of the writer that this will not be necessary.

For many years material requirements planning has been the title of the solution to all the procurement problems in manufacturing. More and more technical manuals have been printed and newer more complex formulas for computing economical order quantities, shop-loading algorithms, and other manufacturing problems examined and solved with MRP. However, many if not the majority of installations remain relatively unsuccessful. In addition, strong resistance to implementation of the system in job shops has created a reasonable question concerning its effectiveness under such conditions. Since at least 50% of all manufacturing firms are job shops, a logical investigation of the value of material requirement planning systems in such an environment is necessary.

Material requirements planning is not new. Road building equipment manufacturers such as Letourneau-Westinghouse and Caterpillar Tractor Company, and in the aircraft industry, North American Aviation used relatively mature systems in the

1940s and 1950s. Even though requirements planning was generated using punched cards. When IBM developed the RAMAC system for disc storage, the bill of material processor, and/or the disc operating bill of material processor (BOMP and DBOMP), these innovations made material requirements planning household words.

As more and more data processing managers subscribed to the system, more and more companies found themselves in the middle of a costly, haphazard, computer-oriented system implementation. Little or no education, training, discipline, or changes to old systems were made before an entire company found itself generating material requirements on a system little understood, on a computer usually too small, and with almost no alteration to peripheral systems to adjust to the new methods. Today anyone preparing to install material requirements planning has some important questions to answer, and the most significant of these is whether the technique is at all logical for use in a job shop—one-of-a-kind—environment.

Figure 159 shows the two "formulas" which emphasize the difference between mass production and a job shop.

Manufacturing companies, predominantly mass production, depend upon forecasts for the various products prepared on some regular basis. Usually this forecast is converted to a "master schedule" which is then used to predict a manufacturing

Figure 159 The basic difference between scheduling in a job shop versus a company mass producing their product is shown here.

"build" schedule and a corresponding cash flow, material acquisition, and production schedule. In the most simple terms, the master schedule or forecast is multiplied by the corresponding bills of material, and total part and assembly requirements are generated on a daily, weekly, or monthly *time-slotted* basis.

During this procedure, part quantities are critical. Economies of scale are made possible by machining or fabricating a large lot size of the same part with one set up. The entire system is part oriented and inventories are "neutral" in that 1000 identical parts may be found in inventory (and required), but not one record showing which customer will finally receive these parts is kept.

On the other hand, job shop orders are often for large units such as boilers, gas turbines, large railroad cranes, supercomputers, large pumps, etc., and often the customer insists on personally inspecting many of the parts which will be assembled into his order. Thus, the paradox of multipart manufacturing versus personal identification of parts begins almost immediately with an order. To make possible such personal requirements, the part numbers of such designs have often been discarded in the system and replaced with job numbers tied to the customer order. Thus, a number such as 1010-7665 may often mean customer order 1010, and job number 7665. Job 7665 and 8823, for example, could be identical parts. The system, however, may well be inflexible and thus unable to compensate for this practice. Of course 1010-7665 and 1222-7665 may or may not be the same part on two different orders.

This seeming chaos has for years been acceptable manufacturing practice in a job shop as many of them made a product where little or no significant competition was present. When competition existed, the system used was normally an equivalent, thus, making all competitors equal. Now job shop costs are climbing, and the engineers and detailers necessary are less available to continually release and redraw or redesign parts and assemblies closely similar to old ones. Nevertheless, there will always be differences between orders in a job shop, and material requirements planning must compensate for these differences.

It is important to look at three different ways to schedule material to be able to compare the methods used in material requirements planning. We need to know if the principles are practical and logical before proceeding.

Figure 160 is a picture of material scheduling often designated *mass release*. The vertical lines denote segregation in time between three major manufacturing functions, ordering, fabrication, and assembly. These categories are still used by some manufacturers.

In the first section — ordering — the horizontal lines represent the procurement lead time of individual parts or raw materials. The term "mass release" is used to describe this system since, upon receipt of an order, every part and raw material known to be required is ordered more or less immediately. In fact, weekly releases of new designs are put in the manufacturing stream, and released to purchasing and manufacturing at once. The gap between these horizontal lines and the next vertical line represents unnecessary wait time where the part or raw material wait for use in the fabrication cycle, thus representing surplus inventory.

The next segment, fabrication, also has horizontal lines representing machining or other fabrication, welding, for example, to make suitable parts. Parts may be manufactured before actually required, therefore creating an additional surplus inventory, but also creating shortages by not manufacturing the correct part at the correct time. It is worthy of note that the less efficient a production schedule is in a manufacturing company, the longer the distance is allowed between the two vertical lines.

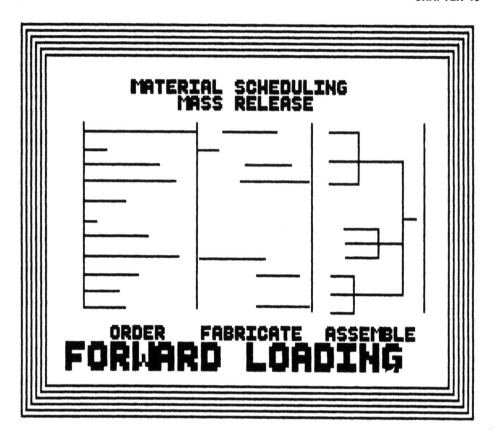

Figure 160   Manufacturing first implemented a production logistical system which can be described as "mass release."

The last segment, assembly, once again represents assemblies of the purchased and fabricated parts, very often with a random schedule applied to them. Usually the parts which are finished first are used to make an assembly regardless of the need. Staging of parts by customer order or by job will radically change the efficiency of assembly. Unfortunately, many concerns wait and stage all the parts before assembly, thereby creating a mountain of inventory.

This entire process is also called *forward loading,* since immediate ordering and designing take place when a customer order is confirmed. An estimate of delivery date is prepared, and is subject to continuous changes as designs, procurement, and fabrication problems develop. The process is costly, but disarmingly simple.

Figure 161 shows a material scheduling system usually called *order point control.* It is a common sense approach developed by manufacturing personnel, who even in a job shop environment, recognized the extent of duplication of parts, assemblies, or raw materials from order to order. Parts which repeated were often called standard parts, and religiously cataloged in books for the engineers to reuse. Kardex systems were devised to keep track of the parts, and traveling requisitions utilized to minimize paper work when reordering purchased parts and raw materials.

Nevertheless, most fabricated parts were presumed to be different from one another from order to order, so most standard parts were purchased parts. Relatively little effort to make designed parts standard was made. In fact it sometimes seems as though the thought of developing a method for retrieval of previously designed parts to avoid repetition is somehow disloyal, treasonable, and, at the very least, destructive of engineering creativity. Thus, order point control held relatively little influence over the majority of inventory created for an order.

Manufacturing techniques changed only slightly. It is true that many parts placed in inventory eased the scheduling and procurement functions, but the scheduling of production functions still maintained the forward loading procedures.

Order point control gained many followers, and it soon became obvious that procurement of parts and raw materials had to be controlled. The order point control procedures examined averaged monthly or weekly usage, safety stock requirements, purchasing lead times, and other factors to determine economic ordering quantities for each item. Unfortunately, several problems crept into this technical solution, since requirements did not always follow the mathematicians desires. Order point control formulas must, by their nature, rely on past history. The readers will recognize how little of the future we can foretell based on past practices and experience. However, order point control had to rely on past requirements, since no easy forecasting method

Figure 161   Order point control was a significant improvement to mass release procedures.

to tie all the parts and assemblies for a product or order was available. The sheer size of the problem made it seem insurmountable.

Thus, order point control of material had many shortcomings. Only with the advent of the computer era and its ability to handle a tremendous volume of data did the problem, finally, seem manageable.

The computer made possible material requirements planning systems. The size of the problem may be dramatized, however, with an example. Suppose we have 100 different products, with an average of 500 parts in each, and wish to forecast the requirement for two years on a weekly basis. The records generated on the computer of quantities alone are 5,000,000 pieces of data. Add to this on-hand inventories, in-process parts, parts on order, lead times, product structures, or bills of material, and suddenly a large quantity of data is required. The first material requirements systems failed. With alarming frequency they are still failing even in the type of manufacturing industries best suited for the procedure.

As shown in Figure 162, in order to use MRP a significant change in the principles of scheduling must take place. For mass release and order point control, forward loading was used. In material requirements planning backward loading is utilized, a 180-degree shift in philosophy. No longer are all or some parts ordered immediately.

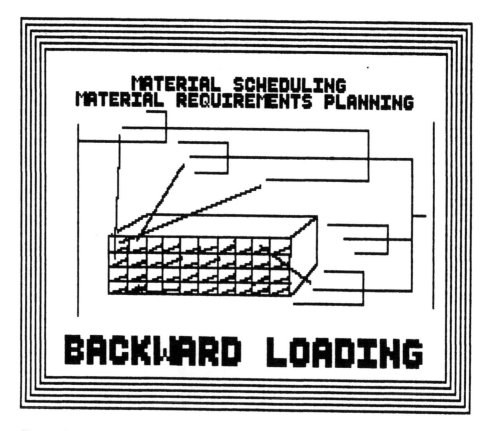

Figure 162 Material requirements planning reverses the entire concept of mass release and also requires a significant increase in the accuracy of data.

Instead the completion date is scheduled, and each assembly and part in the finished product or order is backed off based on assembly, fabrication, or procurement lead times. The entire chain of parts, assemblies, and raw materials for a product is examined as an entity. For mass production, this means of scheduling does not cause a severe problem, because a high quantity of finished products are forecast to a series of delivery dates.

No longer does stock inventory have meaning. All parts and assemblies are in inventory when completed, otherwise the material requirement system continues to call for completed items. Thus, an expanded highly accurate inventory control system, including work-in-process, is mandatory. This includes the need to keep track of all materials on order by time period so extra items are not ordered, and shortages not reported.

Requirements will not be accurate without an accurate bill of material, structured in the sequence of manufacturing, with proper and accurate part numbers to record data correctly. Methods for forecasting must be developed for a master schedule, and accurate lead times created for backward loading and scheduling.

Job shops do not have adequate methods and systems in many of the areas just cited. MRP creates a completely new, rather rigid, and definitely altered business philosophy and climate.

If a solution to job shop scheduling is possible utilizing MRP, then we must attempt to use available techniques and data bases, altering them to fit the job shop problems. The primary and most important data base is engineering documentation, which is captured in the item master and product structure. Figure 163 pictures some data to be found in the engineering file. Normally, specific assemblies are documented in the product structure file. Here assembly part number 1201, is composed of parts 3321, 4267, and 5114. Most assemblies are so documented. Documentation for mass produced products is thus maintained.

Figure 163 shows in addition to this normal engineering file, what we shall call generic bills of material. Several characteristics of these documents are:

1. The product number (700S) is a prefix number.
2. A "pseudo" part number equals product number plus a four-digit suffix.
3. Each part has a generic name.
4. Some known standard parts are included.
5. Standard lead times are included.
6. Generic bills of material are under engineering control.
7. Product structures match how the product will be built.
8. As many generic bills of material as required may be maintained.

Also seen in Figure 163 is a new concept called the *order file*. It is a separate entity from the engineering data base. Here the generic bill of material most closely matching a customer order is converted to the specific order such as customer A order 1010, and customer B, order 1011.

The letter U represents those parts and assemblies which remain generic and are identified with the generic part number only. The letter K represents parts which have been converted to actual designs. Even before actual designs are prepared, simula-

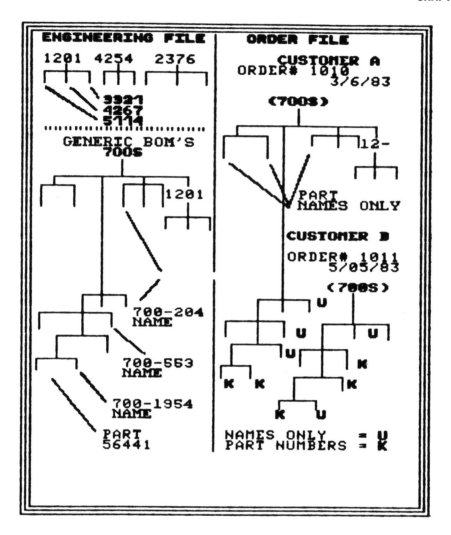

Figure 163 Through the use of product structuring and modular design a generic product structure for a product might be maintained in the data base and transferred to a special order file as needed.

tion of delivery dates may be made using the generic bills of material and their corresponding lead times.

The generic bill of materials now may be thought of as a replacement for a forecast or master schedule. Each generic bill of material may be manipulated individually or in conjunction with all others. The prefix of the part number, formerly the product number may now be replaced with the order number for example, 700S-204 becomes 1010-204, and 1031-204, until part numbers for actual designs replace them.

Examine some of the other features of the order file in Figure 164. It should become the repository of all the specials. The specifications covering the particular order may be maintained as a file attached to the order number. The creation of special instructions such as packaging, personal inspections, heat numbers, and many more may now be tied directly to an order number without fear of confusing the engineering documentation. The bill of material, at first in generic form only, can be converted slowly or rapidly as design progresses.

Another unique feature worthy of consideration is documentation of engineering changes. Often designs are issued prematurely to alert purchasing and manufacturing.

Sufficient data are known to release the part, but final dimensions may undergo several revisions. It is often impossible to know who or how many people need to be informed of each engineering change. By consolidating and summarizing all engineering changes based on each order affected, a clear communication of such changes can be distributed. Less confusion will result. We do not recommend that premature releases be encouraged. The principle set forth is to recognize the reality of design requirements today.

Thus, Figure 164 is an attempt to show how versatile an order file could, and should be to the job shop operator.

Note that invariably a customized order has the engineering design documented from the bottom level of the bill of material first. Under normal MRP considerations, this one fact makes use of "normal" MRP impossible, since accuracy of requirements is based on a complete "chain" of assemblies from the finished product down to the lowliest bolt and nut in the first assembly required. The generic bill of material fills

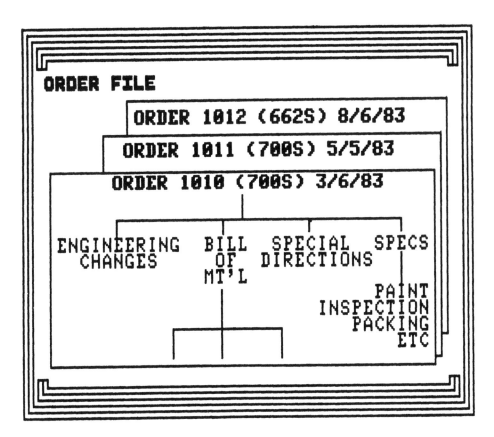

Figure 164   The order file becomes the active data base used to document design as well as material requirements.

this lack of complete documentation, allowing a bill of material to be prepared from the bottom up, not the top down.

Figure 165 shows how the manipulation of engineering changes could occur and be documented. Where each C is shown on the generic bill of material, an engineering change has occurred. The change may affect a generic part number or a "real" part number. It should be emphasized that the great majority of changes requiring communication will be on actual designs.

Each order which has the part number of the part or assembly being changed will keep a summarized record of the changes which occurred during the life of the manufacturing process. It is not unreasonable to suggest that the final bill of material in combination with the record of engineering changes will become the customer's master file, used for spare-part provisioning. It is even possible to consider updating a stored data base for the customer so future spare part orders will automatically route the request to the proper new part number if it has changed.

Since the entire process today of maintaining customer records is usually inaccurate as well as difficult for job shop products, the principle of an order file as described here may more than pay for itself by improved customer service.

Figure 166 pictures the order file in a more dynamic manner. Order 1010 is sched-

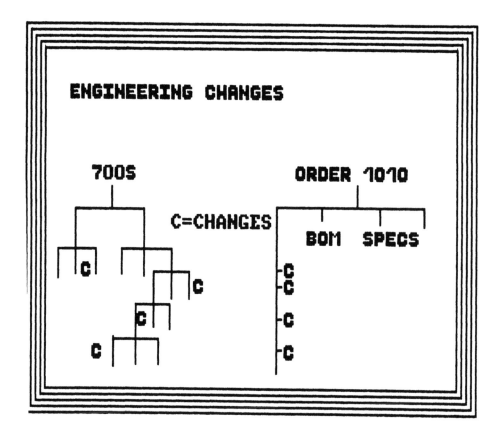

Figure 165   The order file may be used to store specific and customized data about the individual orders.

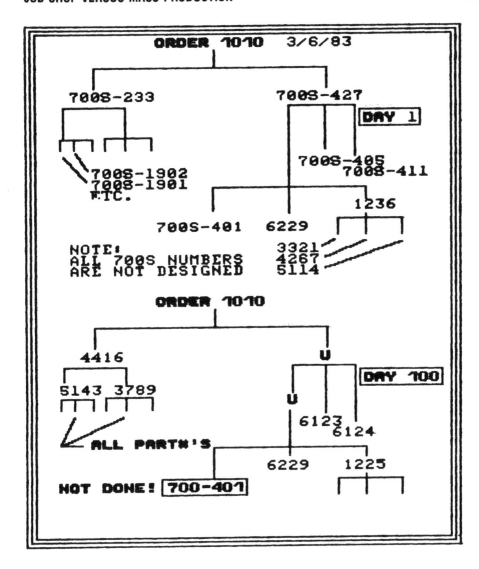

Figure 166 Utilizing the techniques of material requirements planning, the order file data may be manipulated to produce both material and design requirements.

uled for completion on March 6, 1983. On day 1 of the order cycle, most of the parts and assemblies for the order are yet to be designed, as implied by the generic part numbers indicated. Certain existing designs are recorded in the proper place with the bill of material structure as real part numbers. Generic part number 700-401 is used only as an example, and on day 1 of the order represents an unknown design entity.

The same order and its corresponding generic bill of material is shown on day 100 of the order cycle. It is evident that unknown — generic part numbers now have been replaced with actual designs and part numbers. In the illustration, one chain of the bill of material still has generic part numbers (represented by a U) because our example part 700-401 has yet to be converted to a real design.

Since its position on the bill of material implies the need for an early decision of its configuration, it no doubt requires some degree of expediting. If we examine this problem carefully, we are able to see that the principles of the material requirements planning system would have begun this expediting procedure probably on day 1 had we used the customer bill of material in the requirements planning system. The beauty of this idea is that—with no real part numbers—with no certain inventory control—with little formal disciplines—MRP can work to call our attention to possible shortages of material in a job shop environment.

We had best explore this principle still further to see whether the example is an empirical exception or can be extended to the entire material scheduling procedure within a job shop as the general procedure.

Figure 167 shows a representation of the bill of material for order 1010 turned onto its side and extended backward in time from the delivery date of March 6, 1983. This picture is typical of the manipulation of data which occurs in material requirements planning.

On day 1, almost all parts and assemblies for the order show the letter U beside them signifying unknown designs. Only the generic numbers and names of parts exist. The part, example 700-401, appears far to the left on the bill of material. Since

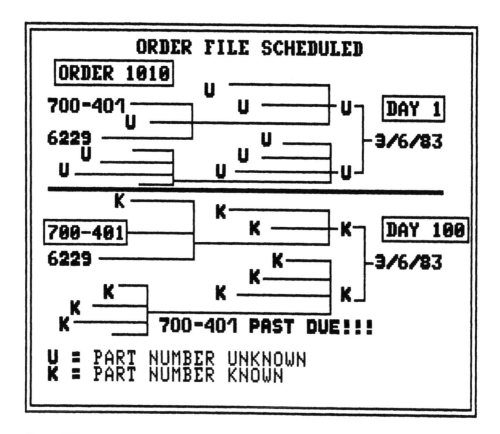

Figure 167 Exploding the bills of material for an order provides a daily, weekly, or monthly record of requirements and provides exception reporting in the same manner as material requirements planning techniques.

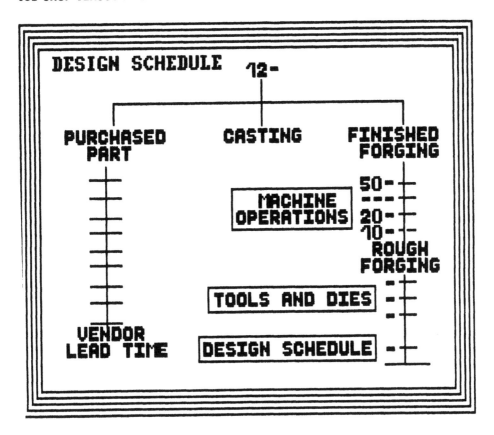

Figure 168 Not only may design and material requirements be scheduled using the order file philosophy, but such special material as forgings, castings, tooling, and purchased parts may be scheduled also.

the extension of the bill of material represents time, it shows that part 700-401 is needed early in the manufacturing cycle.

The second portion of Figure 167 shows the same bill of material extended backwards in time, but on day 100 of the order cycle. Now the majority of the parts have real part numbers as represented by the letter K. However, part 700-401 has not been changed to a real part number and is also past due.

Material requirements planning could have been used to assist in the scheduling of all functions surrounding each of the parts represented on the order bill of material. From the first day of order acceptance, delivery schedules and expediting could have been monitored. Thus, quantities of parts do not necessarily make material requirements practical. Rather, a job shop environment may need material requirements more than the mass production manufacturer.

It seems logical to conclude that the combination of a generic bill of material and a material requirements planning system has real meaning and great significance in a job shop.

We also conclude that there are more capabilities offered by material requirements planning when used in a job shop environment. Let's be sure to look for them.

Figure 168 shows an assembly consisting of a purchased part, a casting, and a forging. The "ladders" below the purchased part and the forging represent time.

Since the vendor lead time can be documented as shown, it can become part of the generic bill of material documentation. In a like manner, the machining operations converting the rough forging to a finished forging may be documented and made a part of the generic bill of material. It is this lead time data which may be used to develop MRP in conjunction with the generic bill of material. Of course, such lead times can change, and it is even more important to alter lead times when the generic parts are replaced with real designs. There is an important addition to these features, however.

Below the title "rough forging" are the titles "tools and dies," and "design schedule." This implies the possibility of adding additional activities serially to the fabrication of the part. Indeed, it is possible to add design time, tool design time, drafting time, and any other requirements which are needed before the actual part fabrication. Thus, an extention to job shop material requirements planning might well be engineering design and drafting scheduling and the tool and die manufacturing cycle.

We have now almost come full circle. In the "old days" production scheduling included such things as design and drafting activities in parallel with material requirements. The technique also had a name, line of balance manufacturing. This mixture of activities with material also has a new name called program evaluation and review technique (PERT) scheduling, or CPM. We can thus recognize that the PERT and CPM techniques, originally designed to manage large construction projects or missile design and fabrication projects are close first cousins to material requirements planning. It seems we have taken the best of all techniques, mixed them with computer systems which already exist, and added a healthy dose of common sense and user participation. Thus, it seems possible and practical to consider material requirements planning in a job shop as a healthy, vigorous, and logical solution to scheduling problems.

Thus, engineering documentation flow is a significant part of and influence on manufacturing systems whether they are in a mass production or job shop. It might even be said that engineering documentation is more important within a job shop than in any other manufacturing atmosphere. So it becomes important to recognize the importance of the rules and principles we have been describing in this book.

THE ENGINEERING
MANUFACTURING PROCESS
11

INTRODUCTION

As shown in Figure 169 the engineering manufacturing process is composed of a series of 26 functions or steps, including those which are emphasized by computer-aided design/computer-aided manufacturing (CAD/CAM). Each function is part of an integrated system. The emphasis on CAD/CAM has created a significant need to understand the entire process described in this book, because CAD/CAM influences and changes each of them. There are many iterations and communication links between them. It is virtually impossible to diagram or explain without developing an almost hopelessly complex picture. The intent, then, is to describe the flow of data or describe the activities or tools necessary for building a good system. Most of the systems outlined are mechanized. Many are also on-line. We believe the interaction of on-line systems will become mandatory for sound management of a manufacturing company.

In defining each function pictorially, as well as literally, some of the major requirements of subsystems needed have also been defined and illustrated. Again it is obvious many variations occur in the principles described here. The intent is to describe one path clearly so that an understanding of how a variation might affect the path can be examined. Many data increments are required to operate an engineering manufacturing system, but only a small portion of them are mentioned specifically or implied by reports. In fact, hundreds of reports possibly are also missing and your favorite may not appear. Brevity has won. Two significant functions will, in the future, determine the effectiveness of the system. These are the design data base, and the engineering manufacturing planning data base. The latter is, in fact, a combination of the design data base, engineering documentation data, and manufacturing planning data. It is, we believe, a dynamic and powerful concept.

Engineering and manufacturing systems are complex but understandable. To emphasize the need for integration, it is appropriate to make a coherent, cohesive schema to picture interrelationships and data flow.

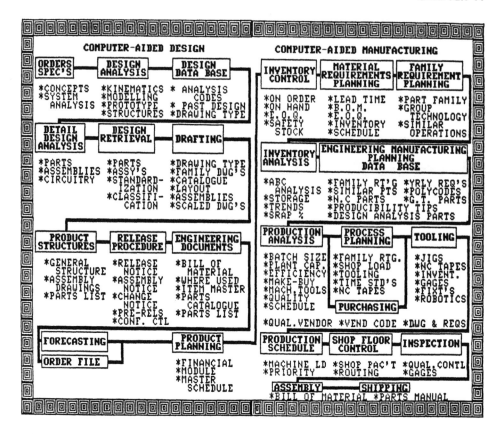

Figure 169 This chart shows 26 basic steps in the engineering-manufacturing process from order to shipping. Divided between engineering and manufacturing, each step also has requirements necessary to fully utilize an integrated system. An integrated CAD/CAM system is well represented using this chart.

ORDERS AND SPECIFICATIONS

Three possible types of activities initiate the engineering manufacturing process. These activities are triggered by an order, a product specification, or general market requirements. Each of these activities cause new designs to be prepared except, perhaps, in highly repetitive products. But of course, these too, are eventually affected by changing customer tastes, or new techniques to perform an identical function, such as electronic chips versus vacuum tubes.

Highly engineered products have a unique set of problems, in that, to some degree, each order is its own prototype. Conceptual design often takes the form of finding the last design, salvaging parts and assemblies suitable for the new order, and then extrapolating the rest of the product to a larger size or capacity equal to the customer's requirements. The experienced product engineer usually has a long history of, and experience in, this "cut-and-paste" approach.

Regardless of the product, it can be enhanced through a systematic analysis of its required function and an orderly examination of the concepts used to perform that function.

For example, a tractor has a series of attachments which can be used to perform various jobs. In addition, it can be used as a "pusher" to enhance a self-propelled earth mover. The simulation of various combinations of equipment, including optimum movement and placement of dirt from an excavation, for example, can be critical to the sale of correct and economical combinations of equipment. Identification of a specific oil source and its viscosity and Btu content is another case of customization when designing a large-scale oil-fired boiler. A standard product may have to operate in a strange and unfriendly environment.

Thus, mathematics, including computer simulation techniques, are infiltrating the marketing and sales departments. Even certification of products for reliability has increased in importance and become a mandatory requirement for such products as hoists, pumps, gas turbines, and, of course, airplanes.

DESIGN ANALYSIS

As specifications or orders are prepared, marketing and, in turn, customer requirements predominate in a free enterprise economy. The lack of this knowledge or interest in it can ultimately ruin a company. However, customer requirements have become more complex and operating characteristics of the finished product more accurate, with ever higher reliability needed. Thus, as we see in Figure 170, it is not only necessary to look at such features as weight, configuration, attachments required, methodology, product life, cost, quality, and style, but also simulate, if not build, a prototype. There is nothing more devastating than designing a product with an ability to be moved from one place to another on the highway using its own propulsion only to find it so heavy that two flatbed trucks are needed to transport it to avoid roadway collapse.

We must therefore be aware that large-scale computers and CAD/CAM have now become significant tools to alleviate these problems with design analysis. It is now theoretically possible to build mathematical prototypes in place of a physical model. The airframe industry has done so for years. With kinematics, three-dimensional geometric modelling, structural analysis, gear train design, and a host of other mathematical disciplines, almost any product imaginable can be simulated. Some of our most complex products, for example, space vehicles, had to be developed in this manner.

Many dynamic simulations in the past used analog computers. It was not easy to take the results of such simulations and integrate them into other engineering design procedures or even compare one result with another. Now the large-scale, high-speed, digital computer has replaced the analog computer. It is fast and economical, and the digital results can be compared and stored, if necessary, for future analysis. Thus, the possibility of a design data base became a reality.

DESIGN DATA BASE

During order processing, or to satisfy the requirements of a product specification, a design data base becomes the key to rapid design and analysis. For example, a new gear box for a standard product may be required to meet a customer requirement.

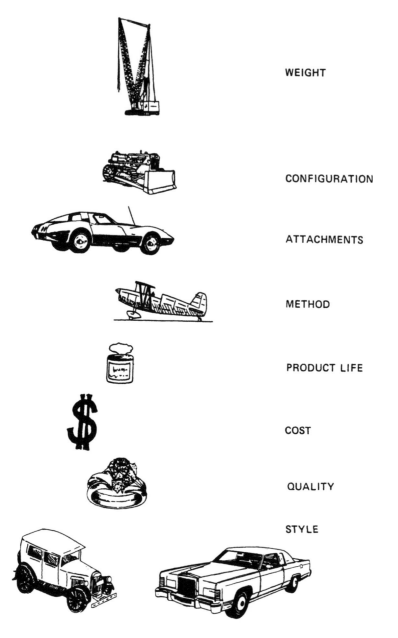

WEIGHT

CONFIGURATION

ATTACHMENTS

METHOD

PRODUCT LIFE

COST

QUALITY

STYLE

Figure 170 Engineering design is not simple. Various marketing requirements, often diverse and conflicting, must be considered as weight, configuration, attachment requirements, design methods, product life, cost, quality, and style. (Courtesy of Control Data Corporation, Minneapolis, Minn.)

Finite element analysis computer codes from the data base may be utilized to engineer the new configuration. This will probably be only one of many design programs in the data base. More and more emphasis will be placed on mathematical models for products and their components. Eventually it will be practical to integrate these mathematical methods, even including empirical formulas. Thus, as modelling and simulation of a complete product is made, the next natural step will be to analyze configuration requirements or changes as an integrated design. If this practice were implemented on even simple designs today, a great deal of design time would be saved.

Finite element analysis is only one of many available design disciplines. Others, as shown in Figure 171, are solenoid design, heat exchangers, rocker arms, cams, linkages, springs, pitman arms, suspension systems, bearings, gears, tube bending, and many others.

There should be no need to include this kind of part as a drafting requirement. Once it has been designed, sufficient mathematical data is available to draw the part automatically. In like manner, some can also be automatically manufactured. If these parts are amenable to manufacture on numerically controlled (NC) machine tools, cutter center-line data can be prepared automatically as well as data to make templates or tapes for NC torch or plasma equipment practical for parts burned from plate.

Finally, confidence in such design programs can eventually lead to a "paperless" factory where design data will be transferred to a direct numerical control computer center in manufacturing which schedules and manufactures the parts.

DETAIL DESIGN ANALYSIS

Inevitably many parts and assemblies must be prepared to make a complete product. The nature of the drawings for these is important and requires definition. Depending upon the product and how it is to be utilized, Figure 172 shows that these drawings may fall into different categories as products, structural, piping, architectural "take-offs", wiring, electronic, purchased parts, or a variety of manufacturing drawings. Knowing the types of drawings and their complexity together with the volume in each category required yearly is very important. Once this data is known, it is also possible to determine drafting and design layout requirements by category plus set productivity standards. Every drawing category can be analyzed to see whether simple two-dimensional (2-D) drawings are sufficient as needed for wiring harnesses, or more complex three-dimensional (3-D) wire-frame drawings as for piece part drawings with three views, or whether geometric modelling, needing surface definition, is required for design analysis or numerically controlled machine tools.

When requisites for each drawing category are defined, standard drafting practices can be applied to all. It is neither wise nor economical to design a part using a geometric modelling system only to have a drafter redraw it to meet drafting standards. The geometric model should be programmed to draw three views of the final design complying with standard drafting practices. Similarly, all like parts which will be machined on numerically controlled machine tools should be dimensioned identically from the same datum lines in order to eliminate redefining the geometry of all subsequent similar parts repeatedly. A classification system, as we will discuss later, will be a most helpful method to assist in this standardization.

SOLENOID DESIGN

HEAT EXCHANGER/TRANSFER

ROCKER ARM CAM

SPRINGS

PITMAN ARM

AUTOMOBILE SUSPENSION

CAMS

BEARINGS

GEARS

TUBE BENDING

CUT AND FILL

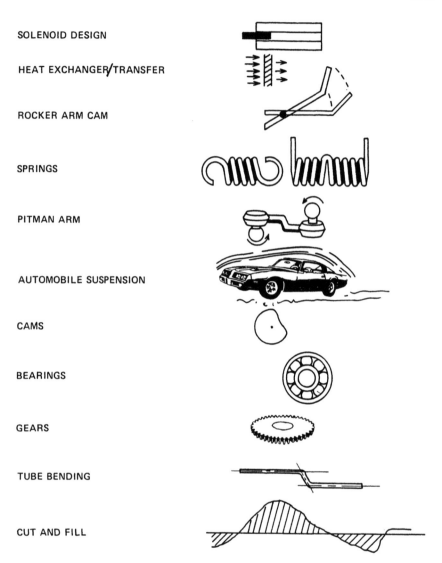

Figure 171   Many other design calculating programs exist, for example, the design of springs, cams, bearings, and more. (Courtesy of Control Data Corporation, Minneapolis, Minn.)

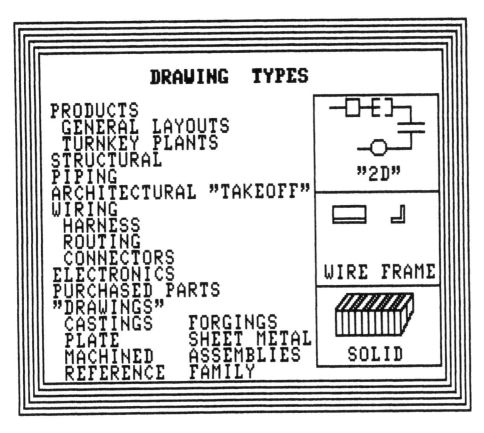

Figure 172 A variety of different type drawings are required in a manufacturing company. Careful analysis of these requirements should be made.

DESIGN RETRIEVAL

Once a product has been designed, or a specific product design change made to meet a customer order, it is necessary to see whether parts and assemblies already exist to meet this requirement. Not enough emphasis has been placed on the need for utilizing existing designs to make new ones. In most manufacturing companies thousands of parts and assemblies already exist, as indicated in Figure 173. Through the years this proliferation has gone unchecked without respect for the cost. Today it may cost from $300–6000 to release a new part before the first machine tool has touched it. Tooling, jigs and fixtures, inspection gauges, templates, routings, inventory stock, spare part provisioning, and many other facets of manufacturing contribute to this cost in addition to the actual drawing time.

In picturing the number of designs available, it is probably almost impossible to easily retrieve a design from the present undisciplined storage to meet a new requirement. It is often said, "it is cheaper to draw a new part than spend all that time looking for an old one." The black books of drafters and designers are of limited and parochial use. Even tabulated drawings fail, because they also have limited use and have been added to with little or no control over variety.

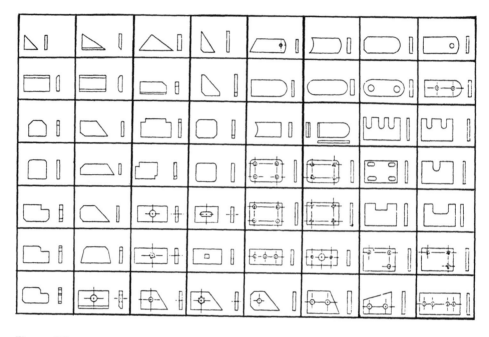

Figure 173 The result of shape coding with a random code number can create a problem. As the number of shapes grow it takes almost as long to find that which is required as looking for the proper design. (Courtesy of American Hoist and Derrick, St. Paul., Minn.)

As an example, one manufacturing firm has a file of 200,000 drawings to which are added 7000 new designs yearly at a cost of $4,000,000. The capitalized value of the drawing file is $114,000,000 and is equal to 28 years of drawing effort. Shouldn't this significant resource be utilized as much as possible? Remember also that when a new drawing is made and used only once, the entire cost must be absorbed on the one order for which it was prepared, increasing the cost of the order.

Finally, a CAD/CAM system will probably increase productivity of drafting by four to one. If 7000 drawings are presently being made yearly, 28,000 drawings can be prepared with CAD/CAM. Where will the manufacturing firm hire the additional industrial engineers to route them and the tool designers to make the necessary tools?

CLASSIFICATION

If some order is to come out of present chaos, then a classification system becomes mandatory. Unfortunately, hundreds of obsolete, inadequate, and unusable systems already exist. Many of these systems were developed years ago, and were indifferently designed with narrow objectives. Certainly many have failed to meet modern needs. The experiences of the past have left a bitter taste in the mouths of engineers and drafters. We cannot afford to repeat past mistakes. No longer should keywords or descriptions be used as a retrieval method. Drawing size is a nonessential part of a code. Making a part number system the carrier of coded information is disruptive to the part number system and creates part numbers which are excessively long.

Rather, the classification system should use permanent characteristics of the parts, whether form, shape, or function to group similar parts into families. Significantly there are characteristics of a good classification system which, when followed, will establish a good retrieval system. The major criteria to judge a classification system is how fast retrieval can be made and the success rate achieved. Too often this is overlooked. As implied by the wheel, the entire universe of a company's parts and other information can be classified into families if the following rules are followed.

The classification system should be:

All embracing

Mutually exclusive

Utilize permanent characteristics

Use logical primary "descriptors"

Tailored to future needs

A predesigned discipline

It is well to remember that a retrieval system also retrieves information as well as designs. Comparison of similar data concerning different parts may well be the most important aspect of the system.

The results and benefits of classification are numerous. First of all, a significant separation of raw materials, designed parts, purchased parts, assemblies and products can be maintained. In the past, emphasis was placed on design retrieval. Now an even more important concept has emerged for similar information retrieval. Once all similar parts can be associated together in families, the proliferation of designs can be examined in an orderly manner and the quantity reduced. For example, the "spigot" family shown in Figure 174 had 27 parts which could be adequately and logically reduced to seven. This example is a microcosm of past design proliferation. Dimensions with no importance varied due only to the unrestrained whim of the designer. Tolerance variations reflected the degree of confidence in manufacturing or the fear of potential product failure. Hole sizes were mostly arbitrary. Who knows the cost for creating two different bolt circles and the therefore unnecessary variety of mating parts, inspection gauges, and test facilities? It is information retrieval as well as design retrieval which makes such studies possible. One firm, Cincinnati Milacron, found that it cost them an average of $4/hole in 1975.

Not shown are the possibilities that parts may have been designed using numerous dimensioning schemes, finishes, plating, and tooling. Only through the association of similar parts can we begin to improve design by comparisons. Cost avoidance for the future will be significant. It should also be remembered that new techniques such as a mechanized bill of material and the associated "where used" data make it relatively simple to find where all these parts are used and quickly determine which of them can be made obsolete and replaced with the proposed preferred parts. Thus, classification performs the following sequence of events:

Proliferation is classified for visibility to compare for optimization to standardize and control

Thus we go from chaos to order back to orderly chaos.

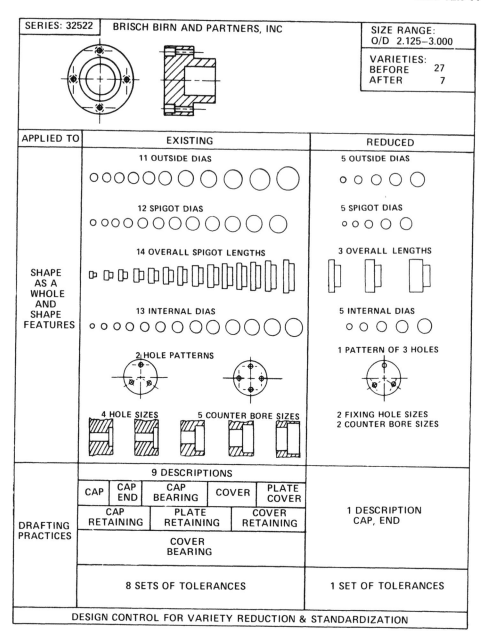

Figure 174 Shown here are the results of a classification system which can bring truly similar parts together so they may be analyzed. Notice how a variety of different attributes of this family could be reduced. The result is 27 parts converted to 7. (Courtesy of Control Data Corporation, Minneapolis, Minn.)

DRAFTING

Once a product specification has been used to create a new product or new parts for an order have been analyzed, the next step is to prepare the necessary drawings of the parts. However, to retain the original goal, the primary effort should be to find parts already designed to meet these new requirements. As an example, a gear box has a round steel bar as part of the assembly plus several parts made from plate. To find the round part family, we follow the branching decision tree prepared for the classification system as shown in Figure 175. Thus, our part is round, single not multiple outer diameter (OD), with a length greater than the OD, with no center hole and no threads. The family of parts thus identified is 30001. In a similar manner, each part required will be investigated.

This inquiry into the retrieval system should begin at the layout stage. It is impractical to draw all the parts required only to find they already exist. Part of the cost for issuing a new part would have then been spent. Unless very stringent rules on the design configuration occur, several parts might meet the design requirement, and be used more easily before the layout is finalized.

It should be noted that all parts in a family may not be "preferred." This is another instance where the designer can be made aware of what direction to take to increase usage of parts already designed, another design economy.

The decision tree pictured here is obviously only a small segment of a classification system. More than 500,000 drawings can be quite adequately placed in families using a five-digit code with a computer properly programmed.

The product of design are drawings of necessary parts and assemblies. If we first draw the parts, or find them, we can later use them to layout the assembly. Of course, the usual practice is to make a layout of the assembly and make details from it. We recommend retrieval of families of similar products, then layouts, then assemblies, parts, purchased items, and finally, raw materials.

However, using the piece parts example, once having identified the families of parts from which to extract the designs, the family drawing may be shown on the screen of a CAD/CAM system. The family drawings are generic in that critical dimensions

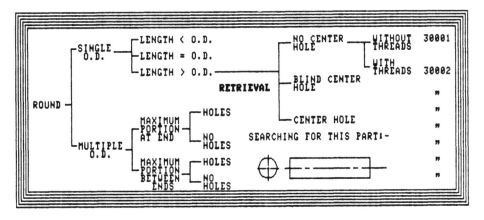

Figure 175 Decision trees like these assist the person retrieving a design to follow a clearly marked and logical path.

are replaced with alpha characters such as, A, B, C. All of the possible variations for the parts in the family may not appear as finishes, plating, notes, etc. However, when a family drawing is identified, an associated catalog may be placed on the screen and examined to see if any parts meet design requirements. Even when one is found and the critical dimensions match requirements, an examination of the actual part will be made to be certain additional variations do not make it unusable. All of these operations should be performed on the same screen rapidly in a CAD/CAM system. Examples of family drawings are shown in Figure 176.

For those part requirements for which previously designed parts cannot be found, a significant productivity enhancement is now available to make new drawings rapid-

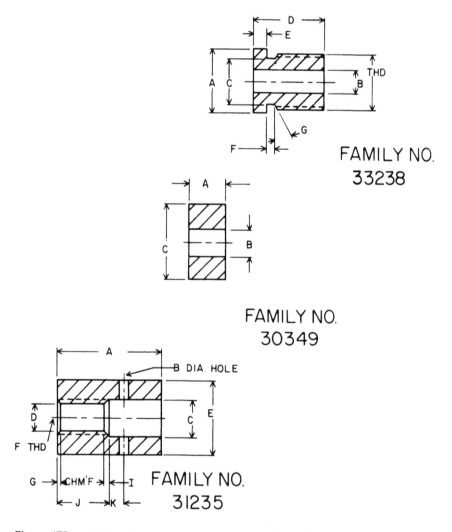

Figure 176   The three drawings represent families of similar parts utilizing shape and form. Each family has a number such as 33238, 30349, etc. The drawing has generic alpha dimensions which can be replaced with actual dimensions when required. (Courtesy of Control Data Corporation, Minneapolis, Minn.)

ly. The family drawing can be used by eliminating and replacing the alpha dimensions with those dimensions actually required. Additional features can be added, including notes, and then a part number issued for the new design. This single feature of a CAD/CAM system tied to family drawings can reduce drafting time significantly, by some estimates, at least 80%.

In designing a new gear box, two new parts have been created, part numbers 421167 and 611277. Previously designed parts have also been found and used, part numbers 711234 and 811423. When properly planned the family drawing will also have a parametric program so that the new parts will also be drawn to scale.

PRODUCT STRUCTURES

Whether from a layout or as a result of direct detailed parts design, it is important to associate the parts in the new design correctly into a "structure." In Figure 177, four specific equal relationships are shown.

The *parent-component* relationship is obvious. However, important communication links between engineering and manufacturing can be shortened if the combinations of parts and assemblies reflect the way manufacturing wants to assemble the product. Thus, a general product structure mutually approved by design engineering and industrial engineering becomes necessary. As companies have two, and sometimes even three, bill of material systems (an extra one for accounting), this creates all the accuracy problems found when two identical systems are run in parallel. A longer lead time is also required to reinitiate data generated for one system and then converted to another. If a list of materials is also found on the drawing, this further complicates the process. Later we will explain how a CAD/CAM system can eliminate much of the paper work and still generate the proper data for the bill of material system.

Pictured beside the general product structure, which is represented by a "Christmas tree" hierarchy of part names, is a specific product structure. Here the names of the parts have been replaced with actual part numbers. As you will see, documentation in this format is not really needed, as a parts list represents the specific structure. It has been shown here to illustrate what is meant by general and specific product structures.

Not shown here is the set of rules for product documentation known as *modular design*. For products with many variables, or attachments and options, modular design techniques further simplify the documentation process, as previously stated.

Having established product structures suitable for engineering, manufacturing, and accounting, documentation of designs is made using an assembly drawing and a "detached" parts list. Each assembly is a *single-level* bill of material in that it is a record of only those parts and assemblies required for a specific *next level*. Thus, if one or more subassemblies are required on an assembly in addition to piece parts, another set of parts lists are used to document the subassemblies.

The list of materials is documented on a parts list and is not included in the drawing. A method must be used to coordinate one with the other. Previously the numbering system was called variously find numbers, balloon numbers, or bubble numbers for this purpose.

The revision level for the assembly represented by the parts list also appears on the document, but not the revision levels of the parts or any subassemblies. Control

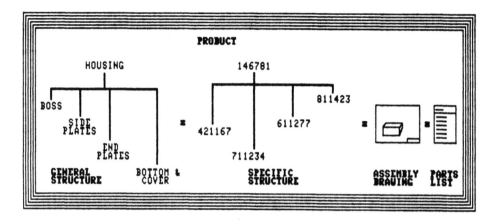

Figure 177   Integration of a coordinated product structuring procedure is essential to an integrated system.

of all documents, particularly when changing a part number or a revision level, is covered under an important set of rules called *configuration control*.

Thus, the parts list is used to document each assembly required for a product bill of material.

RELEASE PROCEDURE

Whether using a mechanized or manual system, it is necessary to disseminate data for the new design to all departments affected, including industrial engineering, production control, inventory control, tooling, and cost accounting. A release notice is the most frequently used method. Each parts list also requires a form to convey the product structure information. In fact, then, the parts list pictured previously was shown prematurely since it is usually only as the design is released, or prereleased that the parts list is prepared. This is particularly true in a mechanized system.

The data thus created is entered into a hierarchical data base to be used to create the item master, and the product structure. A typical list of data elements generated by engineering for each part number and each assembly is shown. The data becomes a significant part of an engineering data base. It is logical to assume it will be combined with part geometry and other design data for a total data base.

In a truly integrated system, the data elements shown in Figure 178 could have been generated and captured while the drawings were being prepared on a CAD/CAM system. The "text" is transmitted electronically to the data base. Thus, the release procedure can be shortened and paper work virtually eliminated by initiating the data immediately as the drawing is prepared. Elimination of such paper work is extremely important because it eliminates an onerous task in engineering. It is time consuming and tedious. To mechanize the work allows more productive and interesting work to be performed by the drafter and designer. It also eliminates duplication of effort which is always prone to error in the transcription.

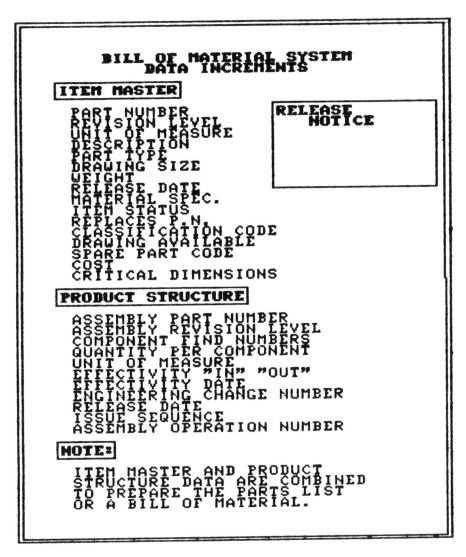

BILL OF MATERIAL SYSTEM DATA INCREMENTS

ITEM MASTER

PART NUMBER
REVISION LEVEL
UNIT OF MEASURE
DESCRIPTION
PART TYPE
DRAWING SIZE
WEIGHT
RELEASE DATE
MATERIAL SPEC.
ITEM STATUS
REPLACES P.N.
CLASSIFICATION CODE
DRAWING AVAILABLE
SPARE PART CODE
COST
CRITICAL DIMENSIONS

RELEASE NOTICE

PRODUCT STRUCTURE

ASSEMBLY PART NUMBER
ASSEMBLY REVISION LEVEL
COMPONENT FIND NUMBERS
QUANTITY PER COMPONENT
UNIT OF MEASURE
EFFECTIVITY "IN" "OUT"
EFFECTIVITY DATE
ENGINEERING CHANGE NUMBER
RELEASE DATE
ISSUE SEQUENCE
ASSEMBLY OPERATION NUMBER

NOTE:

ITEM MASTER AND PRODUCT
STRUCTURE DATA ARE COMBINED
TO PREPARE THE PARTS LIST
OR A BILL OF MATERIAL.

Figure 178 A mechanized bill of material has two associated "files" in the data base, the item master, and the product structure. Merging these two files produces many documents without the necessity of creating redundant data. Some of the usual data increment found in the system are shown.

ENGINEERING DOCUMENTS

When properly structured, the result of preparing parts lists for a product is the capability to generate a bill of material automatically. The inverse of a bill of material is a *used on,* or *where used* listing by part number. This, too, can be generated automatically. Some examples are shown in Figure 179.

Many forms of this output can be prepared. Later we will see how picking lists are prepared from the system. Summarized bills of material, an order-oriented bill of material, the item master file data, and a similar-parts catalog are a few additional examples. The bill of material in Figure 180 is known as an indented bill of material, since as each subassembly and/or parts for the next lower level are listed, they are also indented on the page to show the hierarchical relationship. Familiarity with this report makes it rather easy to determine how a product is assembled. The indented bill of material is also a tool for cost accounting. It allows standard, actual, or average costs for each part and assembly to be "rolled up" to obtain a product cost.

Many job shop (or order-oriented manufacturing companies) do not make proper use of a bill of material system. They cite the fact that part numbers are still being issued for new designs as the product is being shipped. How then can it be used to issue parts or cost a product? A relatively new concept is to create a generic bill of material for the product, using dummy part numbers, and maintaining it in the engineering data base. When an order is received, the generic bill of material is then trans-

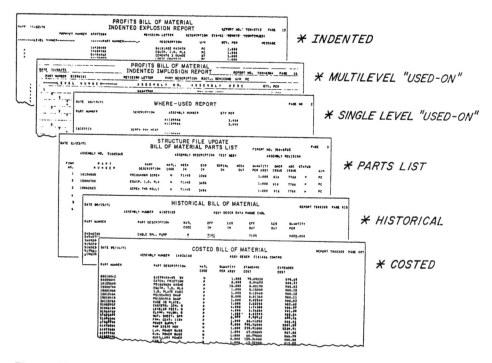

Figure 179 Some of the actual printed output documents obtained from a bill of material system are shown here.

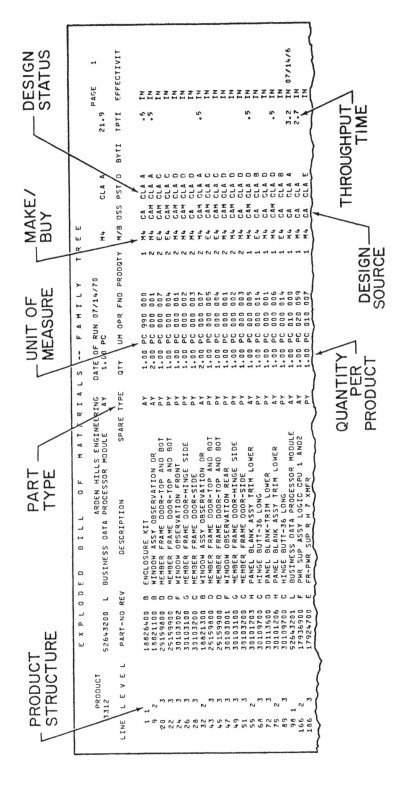

Figure 180 The structural, or indented, bill of material is more clear when printed as shown here. The body of the information is not indented. Columns are provided not only for a level number, but the level numbers are indented to show the relationship.

267

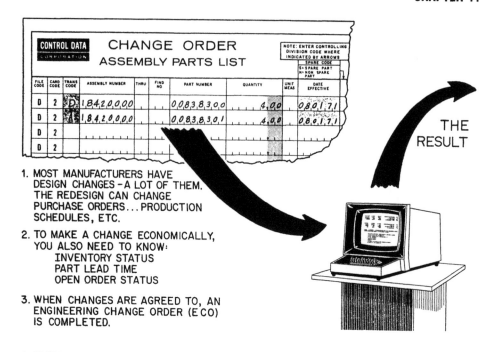

1. MOST MANUFACTURERS HAVE DESIGN CHANGES – A LOT OF THEM. THE REDESIGN CAN CHANGE PURCHASE ORDERS... PRODUCTION SCHEDULES, ETC.

2. TO MAKE A CHANGE ECONOMICALLY, YOU ALSO NEED TO KNOW:
   INVENTORY STATUS
   PART LEAD TIME
   OPEN ORDER STATUS

3. WHEN CHANGES ARE AGREED TO, AN ENGINEERING CHANGE ORDER (ECO) IS COMPLETED.

4. THE PARTS LIST AND BILL OF MATERIAL REFLECT THE CHANGES.

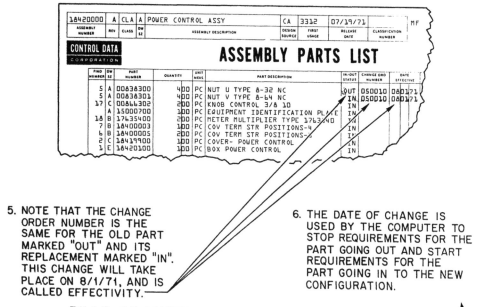

5. NOTE THAT THE CHANGE ORDER NUMBER IS THE SAME FOR THE OLD PART MARKED "OUT" AND ITS REPLACEMENT MARKED "IN". THIS CHANGE WILL TAKE PLACE ON 8/1/71, AND IS CALLED EFFECTIVITY.

6. THE DATE OF CHANGE IS USED BY THE COMPUTER TO STOP REQUIREMENTS FOR THE PART GOING OUT AND START REQUIREMENTS FOR THE PART GOING IN TO THE NEW CONFIGURATION.

BILLS OF MATERIAL CAN ALSO BE GENERATED

Figure 181 The engineering change notice is the document used to change the configuration of the bill of material. The effective change is shown (both the "in" and the "out"), with the date of effectivity.

ferred to an order file, where it is updated as needed by replacing the dummy part numbers with actual. If lead times are added to the generic bill of material, it can be "exploded" (as will be explained later when we discuss material requirements planning) just as any other bill of material used in mass production manufacturing, and requirements for parts generated by time period. This will be a major breakthrough for one-of-a-kind engineered products.

The only constant part of engineering design is engineering changes. They are expensive, time consuming, and retard manufacturing processes. Consequently, a formal change notice procedure is mandatory. A CAD/CAM system must be designed to allow tracking of these changes. Since an engineering change can affect every department of a company, the documents and parts affected must be meticulously documented and the change communicated. It is also important to record the reason for changes into categories to evaluate the quantity and see why they are made. Often a change order board is instituted in a company composed of members from engineering, industrial engineering, production control, and accounting to review significant changes. Severity of change, when documented, may assist in developing a priority system as well as a distribution system, so that those people not involved will not receive unnecessary documents. Figure 181 shows both a change notice to an assembly and the result of the change on the corresponding parts list.

A continuous conflict between engineering and manufacturing documentation can be resolved using a system called *effectivity*. When a designer makes a change to a drawing, it is, as far as he or she is concerned, made immediately. Many changes affect parts only. Many, however, affect assemblies. In the example shown, part number 00838300 has been replaced with part number 00838301. If effectivity is not used, the old part number would disappear immediately from the new parts list. Thus, if the same data is used for a picking list in manufacturing, the new part will be prematurely issued. For nonmandatory changes this becomes disastrous, as a large inventory of old parts may be available. Before effectivity was implemented, this problem made both an engineering and a manufacturing bill of material necessary. On the parts list shown, both configurations appear, and the effective date of change is documented as 8/1/71. Hence, the best of both worlds are available, the new design, and the old one. For the picking list, the computer chooses the proper part based on the effective date.

FORECASTING

It is not readily apparent today how CAD/CAM may affect forecasting. Many job shops and companies with highly engineered customized products can't, or won't, forecast. For mass production products intricate mathematical models of the market place have been prepared. Not only are internal statistics used to predict what must be manufactured based on forecasted shipments, backlog, and inventory, but also extrinsic factors closely related to the product are manipulated in such techniques as multiple regression analysis, exponential smoothing, and triple-exponential smoothing to name a few. Those companies with an effective forecasting model rarely advertise it. Future success in the market depends upon a competitive advantage. Figure 182 shows some of the statistics essential to assist in the preparation of market forecasting.

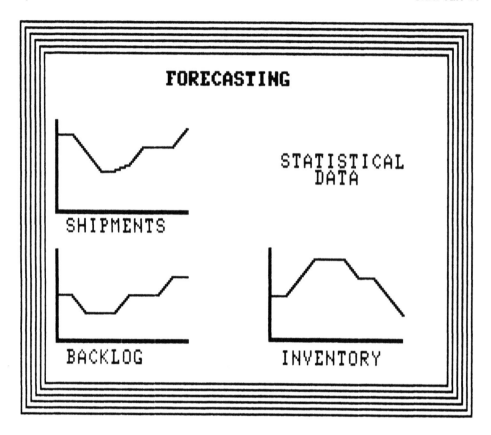

Figure 182 Forecasting of product requirements is an important tool for a manufacturing company. The forecast not only drives the manufacturing systems, it is used for cash flow and capital requirements.

Rare is the condition where any scientific method is used to determine what effect new and altered designs, new attachments and options, or new methodology will have on the company's market. Now it is possible to have a complete mathematical model of a product and the ability to use it for simulations under actual operating conditions. The ease of change that such a model allows may have a singular effect both on design changes and the market.

For those companies with little or no forecasting capabilities, or with highly engineered products, the order file can be used as a "mini" forecasting system. By entering a variety of product mixes based on potential orders, the generic bills of material can be used to determine long lead time items, at least, and can be used also to compute a rough cut shop load, cash flow, and other statistics. Once an engineering manufacturing planning data base is prepared, no matter how rough, plans with an increased degree of integrity can be made.

To be effective, the forecast must be continually monitored to reflect the dynamics of change in the market.

PRODUCT PLANNING

As a forecast is prepared, it must be converted into some form of financial model. At this level of planning, only gross product requirements may be needed. The financial model is a document showing, by time period, the finished products to be built. The manufacturing cycle plays an important role in preparing the model. As already mentioned, it is often almost impossible to forecast products built to special order. The manufacturing cycle is long. An oil-fired boiler, for example, may take five years to design and build. PERT charts have often been used to document the design and construction of large "turnkey" plants.

Many products including tractors, hoists, and large pumps were formerly built to stock. Today the excessive cost of carrying inventory has almost eliminated this luxury. Therefore, only consumer products such as refrigerators, irons, toasters, and the like can be reasonably considered finished goods. Figure 183 indicates the difference in lead time and production time between various types of products.

This set of problems has created the need for modular design. By separating the "constants" from the "variables" in a product, it is possible to generate bills of mate-

Figure 183 Lead time is an important consideration in obtaining and maintaining customer satisfaction. Modern tools can be used to shorten lead time even in a job shop.

rial for portions of a product that have a high degree of repetition in orders. Also, those portions of a product which are self-contained but highly variable such as pump impellors, transmissions, and engines are classified as attachments, and percentages of each variation are forecast separately. Those parts of a product which would be "nice to have" are classified as special options and often are not forecast at all, or only sporadically.

Thus, marketing and forecasting create additional needs and superimpose their special problems on engineering and manufacturing.

A financial forecast is useful for many departments of a company. It can be used to predict profit-and-loss statements, cash flow, purchasing commitments, manpower, departmental budgets, and production and machine tool requirements. In fact, financial planning has also become a highly sophisticated function in many companies and is often used to simulate the effect, for example, that a labor contract will have on pricing or profit. Some companies have utilized this approach to simulate the effect their competitors will have, both on the market place and on their own financial planning. These mathematical econometric models are closely guarded secrets.

A financial model is almost meaningless to manufacturing. It must be converted from general products to actual part numbers. It is this close coordination which requires the price manuals and sales brochures to be parallel with the product designed. The engineering bill of material must also be structured as the product will be sold. Modular designs are ideal to meet this requirement. The module forecast can be considered a direct conversion from a financial forecast to engineering part numbers. Once again, these requirements are extended through time periods.

The last schedule required is called the master schedule. These requirements have been "leveled" to eliminate sharp increases or decreases in the manufacturing cycle. It is a rule that this leveling procedure may never show a requirement less than forecast. Thus, several months may show a schedule of modules higher than forecast so that the modules will be available for months with higher requirements.

This series of forecasts has often been called an *executive schedule* and is approved by the officers of the company. The master schedule has also been called the *build* schedule. The latter is also the trigger starting all manufacturing functions needed to produce a product.

INVENTORY CONTROL

If any division between CAD and CAM exists, it begins with inventory control. All functions defined thus far are those which build data about a product, and could be considered static. Action data is relatively small. With inventory control, dynamic data is being generated which requires constant decisions and action. A mechanized stock inventory has become commonplace. More controls and higher accuracy, including a secure warehouse, have made it possible to control inventory balances effectively. Material requirements planning has made it necessary for even more stringent control. Not only must work-in-process inventory be accurate, but on hand balances of parts are insufficient. It is now not only necessary to know balances and parts on order, but also the time period in which these parts will be received. Economic order quantity calculations are necessary to find the optimum between excessive set-up cost in manufacturing and the cost of carrying surplus inventory. Other methods of controlling inventory such as *part-period balancing* are also being instituted.

Figure 184 shows a series of reports with some of the data necessary for perpetual inventory control. The quantity of parts on hand and on order are shown, and other reports predict when the parts on order will be received. The economic order quantity for each part is shown. The safety stock quantity is a fraction of previous order point quantities. When withdrawal from inventory fell below the safety amount, an order was generated. Many companies now fold the safety stock into the forecast requirements, thus, reusing it several times during the year.

Inventory management is becoming more scientific. With such tools as cycle counting, weekly audits of transactions, usage analysis, audit trails, stock location systems, and even electronic storage and picking equipment, companies are getting closer and closer to optimum inventories.

MATERIAL REQUIREMENTS PLANNING

The material requirement planning system has not been accepted or successfully implemented by a large number of companies. The necessary disciplines to make MRP work have been inadequately explained and poorly planned. It is time facts were separated from misinformation about the technique. It is the most efficient method for planning material acquisition. Even a company with engineered products will benefit. Like so many tools, if abused, it will not perform!

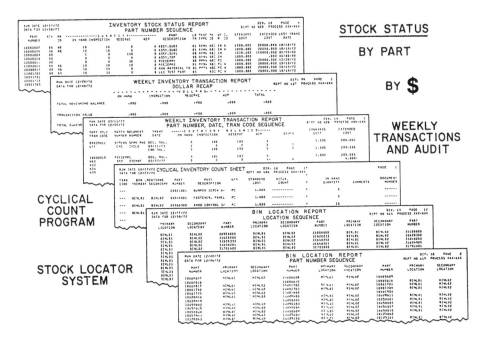

Figure 184 Inventory management not only is mandatory for material requirements planning, analysis of inventory can become an important management tool.

A brief explanation of alternates may put MRP into perspective, and these are shown in Figure 189. The oldest method for providing materials to manufacturing is called *mass release*. Three general lead times for a product were defined: (1) material acquisition, (2) machining, and (3) assembly. When an order was made final, all materials and parts were ordered as quickly as possible. An informal procedure created some priority so that forgings, castings, and special parts were ordered first. However, many parts with short lead times came in long before they were required. In like manner, fabrication began, and once more, machined parts and weldments were prepared prematurely. The final assembly activity was based on staging of all parts necessary before assembly began. Many problems in sheer logistics occurred, not the least being the "borrowing" of parts from one job for another. This kept expeditors in a

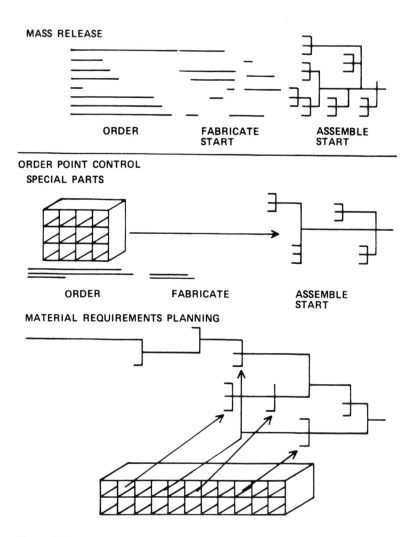

Figure 185 Material scheduling has been modernized from simple "mass release," to material requirements planning using the computer.

continual stage of crisis. The system was called forward loading, as previously mentioned.

Through the years manufacturing saw that different jobs required the same common parts. These became the stock parts we know today. The parts, however, were not necessarily the most important ones, or those with the longest lead times. Nevertheless, an inventory allowed purchasing and production control to concentrate their scheduling activity on fewer parts. Since it wasn't always known where stock parts were to be used, a system called order point control was initiated to determine when to reorder stock parts. This was and still is inaccurate since it is based on past part usage, a poor criteria. All the problems found in mass release systems still prevailed.

Material requirements planning is 180 degrees from the principles of mass release. It is described as *backward loading*.

In material requirements planning, instead of releasing all orders for material as rapidly as possible when an order is placed, a forecasted finished product date is designated. Using the bill of material system and attaching lead times to each part and assembly, the bill of material is exploded backwards from the set completion date. Each part and assembly is "slotted" into a completion date of its own. The explosion is analogus to vector analysis used to determine stresses in some structures. The load in KIPS for each member can be diagrammed with each vector length equal to the force or reaction. In the case of MRP, the vector is lead time.

It is apparent that orders for material can flow more smoothly and scientifically (if lead times are correct). Efforts of the purchasing and production control departments are more balanced. Staging is modular because the completion of each assembly is now time-oriented and will be placed in inventory waiting for the next level of assembly. Inventory control must be very accurate. Rules to become a class A user of MRP have been developed.

Material requirements planning is not new. The procedure has been used by companies such as Letourneau Westinghouse since 1940. Automotive and tractor manufacturers have been active in its use since at least 1950, as well as some airframe manufacturers. Computer manufacturers would have been in big trouble without the technique. Originally, punched-card bills of material were used as the data base. Thousands of cards, awkward to manipulate, were prone to errors. Advanced, large-scale computers with large volumes of random-access memory now make MRP work effectively.

Material requirements planning performs a complex function. Picture a company with 100 products, each with 5000 parts and assemblies. Requirements for each product for 52 weeks may generate 25,000,000 records. If each month or each week product requirements change, these records are "regenerated." Regenerative requirements planning consumes much computer time, and, if the bill of material is inaccurate, errors are almost impossible to find. Talk of these problems caused many people to avoid using the system.

During the past few years two new facets of MRP have alleviated many fears of users and also eliminated the shortcomings of MRP. "Pegged" requirements and "net change" procedures have shortened the computing time and made tracking errors fast and easy.

When the bill of material for a product is exploded and requirements generated, inventory on hand and on order are automatically subtracted. At the same time, a *double-entry* system of demands and supplies is maintained. On the first record, the demand for the product is shown. With no economic order quantity (EOQ) and a lead time of one week, the supply requirements appear on the supply line one week earlier. These supply orders are used to generate demand orders for the next level of parts. If the lead time for this part is three weeks, the demands are backed off by three weeks. The EOQ for the part may show a quantity of 40, and so supply orders are combined into block orders of this size. Thus, level-by-level through the bill of material demand and supply orders are generated. It can be seen that the part number for which the demand order is generated can be identified at all times, and pegged requirements are possible. By maintaining in storage a continuous record of demands also called gross requirements, it is not necessary to change all records in the system, only those affected by some product requirement change, net change of requirements results.

Forecasted spare part requirements can be superimposed on MRP and demands for them kept separately. Accuracy and good auditing are becoming commonplace using an MRP system.

We have described a complex system which is a five-dimensional array of data. It could not be manipulated manually in a timely or economical manner. A computer is the only tool capable of this. The five dimensions we have described are product requirements by time period, product mix in each time period, levels in the bill of material, lead time, and a low level code. The latter is a coding method which is used to explode the bill of material rapidly.

FAMILY REQUIREMENTS PLANNING

A fascinating development using MRP is apparent if we use a classification system in conjunction with it. As mentioned before, a product forecast is extended through time for 52 weeks or more. Thus, requirements for all parts and assemblies are also extended through the same period. As manufacturing begins to plan production, the quantity of a part required is studied to determine whether several weeks production can be manufactured together in one lot and thus avoid several costly set ups. Economic order quantity formulas are used to determine whether it is economical to so plan. Several weeks of requirements may be grouped together. Even though proven more economical, surplus inventory is the final result.

Group technology, which will be described later, has been an alternate technique used to group similar parts in "batches" to avoid excessive setups. However, the method has not been applied at the source of requirements generation or MRP. Group technology also applies to parts with small lot requirements.

Figure 186 shows part requirements by time period, but grouped together by classification families. With this information, requirements of similar parts can be summarized vertically rather than horizontally by time. It will then be possible to both avoid excessive set-up times and excessive inventory. By grouping similar parts needed each week manufacturing by weekly lot can be considered, much the same way group technology is defined. However, all the potential requirements may now be examined, combined, and entered into the production schedule ahead of shop floor control.

An additional advantage of this method may well be the elimination of many shop orders and detailed labor reports.

INVENTORY ANALYSIS

Too often companies spend vast sums of money designing and implementing systems without taking full advantage of the data generated. One example is the use of an ABC inventory report. Figure 187 is an example of a list of all parts manufactured and their yearly usage. The cost for each part is also shown. By multiplying the yearly usage by the cost, the total dollars of usage for each part can be computed. If the usages are listed by total dollar, from high value to low, and the percentage of each value compared to the total value for all parts, the listing shown here will be the result. The ABC inventory formula is based on Pareto's law, which explains the frequency of statistical events of this type. It has also been found, with almost standard frequency, that 80% of all money spent is on 5% of the part numbers used. Another 15% of all money is spent on 30% of the parts, while only 5% of all money is spent on 65% of all parts. It is clear that these break points can be used for many studies.

Obviously, close control of A parts is required and may well be purchased or manufactured discretely. B parts are those most probably purchased using an economical order quantity formula, while C parts can be stored as floor stock, or maintained using a *two-bin* inventory system.

Notice, however, that manufactured C parts may be the best candidates for group technology, A parts may be the best candidates for raw material standardization, and certainly should be relatively free from engineering changes. Tooling, fixtures, application of numerically controlled machine tools are other facets for study using this division.

Figure 187 represents the dollar volumes of an ABC inventory analysis, and another important area for investigation is revealed. Most ABC analyses take past usage to determine the dollar volumes. What would happen should the forecasted product requirements be used?

Suppose the master schedule were to be exploded so that, for each part, total requirements for the year are developed. If inventory on hand and on order are also compared to these total requirements, and the entire set of data is listed as for an ABC analysis, what could be determined from the report?

We are able, first, to study a future and current ABC analysis. Parts whose demands are increasing or decreasing will take their proper place in sequence. The A parts may cover only two or three pages. One company found a startling surplus of several parts forecast for the year in the A category on the first page. Negotiations for yearly purchase order contracts, for new machine tool requirements, for coordinating buyers and vendors will also result.

If all parts in inventory are made a portion of the list, not only will excessive inventory of current parts appear, but all parts in inventory for which there will be no requirements. Surplus inventory will no longer be guesswork.

We have discussed only a few of the possibilities of analysis. Let us combine even more tools so that analysis can be made of manufacturing facilities, and manufacturing methods.

Since many studies of a manufacturing facility had to be performed manually, such studies could only encompass one or two variables. Data was (and is) hard to find. Now that large-scale computers and data bases are readily and economically available, we can shed the manual shackles.

FAMILY PLANNING

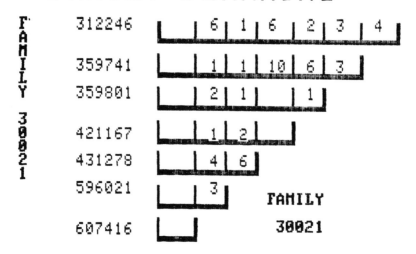

FAMILY 30021

312246	6	1	6	2	3	4
359741	1	1	10	6	3	
359801	2	1		1		
421167	1	2				
431278	4	6				
596021	3					
607416						

FAMILY

30021

FAMILY PLANNING

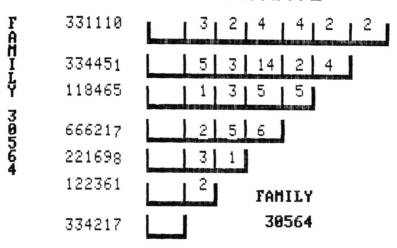

FAMILY 30564

331110	3	2	4	4	2	2
334451	5	3	14	2	4	
118465	1	3	5	5		
666217	2	5	6			
221698	3	1				
122361	2					
334217						

FAMILY

30564

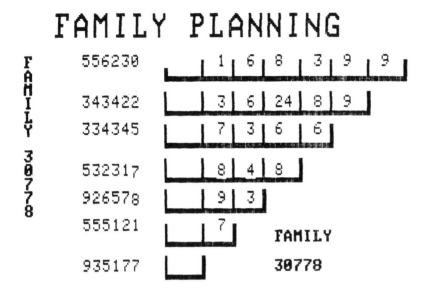

Figure 186 Today if we were to utilize material requirements planning as usual but combine similar parts by family, short runs of parts could be increased by using a form of family group technology.

As prepared in Figure 188, let's use the ABC analysis again, but format it in such a way that all parts are sorted into families by the classification system. In addition, we will look at all parts not only by volumes of money, but also by volumes of pieces required. The array pictured here is an attempt to picture this combination.

With this data we can now superimpose or compute manufacturing requirements. Each family of parts requires one or more machining operations, for instance, drilling, milling, and/or boring. Later we will discuss polycodes which can be specifically used to define machine tool operations. Each part family also has requirements for turning, keyways, splines, and plating to name a few.

We are now able to find our way through a bewildering quantity of data and determine several major facts:

The quantity of machine operations

The type and quantity of part operations requirements

The parts most economical for a dedicated machine line

The parts most conducive for group technology by cells, or production units

The parts most conducive for group technology by identical operations

Total plant capacity requirements by machine

Parts by type and volume for NC tools

Numerically controlled machine

Tool usage and load

DOLLARS

TYPE

Item No.	Classification	Annual Units	Unit Cost	Annual Usage	Cumulative $ Sales	%
683246	37712-004	51 553	317	77 102	1 652 385	5 0
703611	36115-002	98 406	470	46 251	3 304 769	10 0
965527	36223-001	6 768	4 876	33 001	4 957 154	15 0
932016	36112-002	4 250	7 369	31 318	5 254 533	15 9
894642	34117-002	44 560	675	30 078	5 618 107	17 0
537231	34220-001	8 680	3 286	28 522	5 882 489	17 8
202646	34220-002	27 581	930	25 650	6 609 538	20 0
162678	36620-001	3 428	5 900	20 228	7 600 969	23 0
198155	36711-004	52 765	368	19 993	8 261 923	25 0
928234	36777-001	1 105	14 676	16 217	9 914 307	33 0
856549	35002-001	23 908	640	15 001	10 443 070	31 6
903432	35002-010	2 690	5 475	14 723	11 004 881	33 3
910211	35007-019	11 378	980	11 150	13 219 076	40 0
567809	31001-001	244 690	045	11 011	13 252 124	40 1
933986	30110-002	22 224	450	10 001	14 276 602	43 2
910932	30349-002	7 391	054	7 790	16 523 345	50 0
262011	30776-001	2 089	3 540	7 396	17 184 799	52 0
525105	31876-001	56 304	115	6 475	18 209 277	55 1
786399	31677-001	9 984	556	5 551	19 828 614	60 0
584348	33238-005	3 756	234	4 635	21 414 903	64 8
376242	32100-001	21 683	205	4 445	21 844 523	66 1
563453	32101-001	23 796	181	4 307	22 042 809	66 7
579996	32102-002	33 743	113	3 813	23 133 393	70 0
612172	32103-002	7 239	490	3 547	23 662 146	71 6
201808	32104-003	3 571	840	3 000	25 050 149	75 8
998699	32105-003	14 774	190	2 807	25 413 674	76 9
662163	33238-002	1 500	660	2 175	26 133 117	80 0
237495	31235-009	1 212	1 876	2 274	26 834 724	81 2
350198	31295-001	9 967	209	2 083	27 429 583	81 1
264353	31296-001	1 430	1 720	1 913	28 603 105	82 2
782245	33455-001	3 509	450	1 579	29 015 872	86 0
463812	33456-001	243	5 729	1 391	29 445 492	87 0
217486	33238-001	1 942	1 256	1 309	29 742 921	87 1
590427	33567-001	2 392	475	1 112	30 403 875	87 9
260184	33568-001	2 857	250	1 020	30 536 066	89 2
621958	33569-001	15 360	064	985	31 392 390	90 2
206874	36789-001	3 494	156	614	31 891 021	96 2
741326	33333-001	1 904	282	537	32 122 855	97 2
377241	33334-001	2 842	120	341	32 618 072	98 7
977369	33335-001	2 439	123	300	32 717 213	99 0
661393	33336-001	2 670	103	275	32 783 308	99 2
261853	33337-001	3 750	048	180	32 915 499	99 6
2605	33338-001	198	505	100	32 998 118	99 85
6562	33339-001	210	143	30	33 034 477	99 96
613216	33310-001	500	620	0	33 034 781	100 0
3742	33311-001	1000	20	0	33 034 981	100 0

80%   5% A

15%   30% B

5%   65% C

Figure 187 When analyzed, the usage listing shows roughly 80% of all dollars are spent on only 5% of the parts, 15% of all dollars on 30%, and 5% of all dollars on 65%. (These percentages may vary from company to company.) This distribution is called an ABC inventory analysis. (Courtesy of Control Data Corporation, Minneapolis, Minn.)

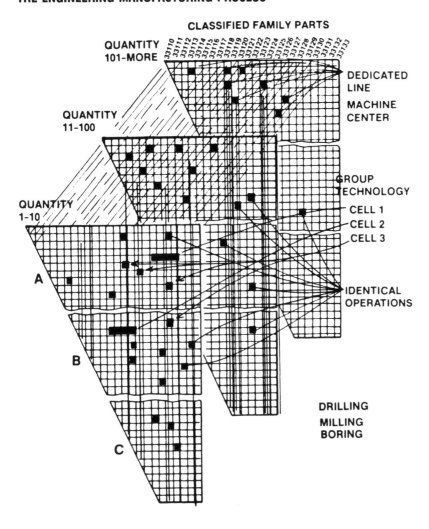

Figure 188 The production analysis will show those parts which can and should be massed produced, those which may be manufactured using group technology of cells, and those utilizing group technology for identical operations. If each part has a manufacturing polycode tied to it, providing machining characteristics, machine capacity, and requirements may also be measured. (Courtesy of Control Data Corporation, Minneapolis, Minn.)

The time for large-scale scientific studies of plants and facilities is at hand. We must take advantage of large data bases to make studies for better and higher productivity.

ENGINEERING MANUFACTURING PLANNING DATA BASE

We have now reached a significant plateau. The engineering manufacturing planning data base combines completely the power of both CAD and CAM, making it CADCAM without a slash. We have taken all the powerful data prepared in engineering and marketing, and superimposed upon it action data prepared from inventory control and material requirements planning. We can see clearly now what we once glimpsed only a shadow of.

In one startling snapshot we can see an entire universe of parts captured into logical families of similar parts. We know:

The total number of parts in each family

How to standardize these parts

The preferred parts in each family

The type of drawing we must prepare

The family drawing we can use for each part

Which parts must be analyzed by finite element analysis

Which parts have other design analysis programs and thus drawn automatically

Which parts will be scheduled on NC tools

The family routings available per family for each new part

Which parts are best for group technology

Total yearly requirement of all parts by part number and family

We are free to manage our design and manufacturing on a flexible, knowledgeable, and economical basis. People, money, and machines can be balanced with foresight. We are truly in control as we have always wished to be.

PRODUCTION ANALYSIS

We mentioned previously the vast amount of money spent to develop systems without fully utilizing the data generated. In the same manner we described inventory analysis methods, production can also be analyzed. The data base we have just described indicated several tools for production. By entering the data base in an orderly manner, we cannot only seek family part geometry and other engineering data as shown in Figure 189, but also family processes, NC cutter center-line data such as a family input trace, Industrial engineering standards, and producibility tips.

Family processes can be used to minimize various paths through manufacturing shown on the processes and choose the best. Set-up times can be standardized. Machine tool utilization can be optimized. If, coupled with them, a picture on a CAD/CAM system is prepared showing best methods for dimensioning, tolerancing, notes,

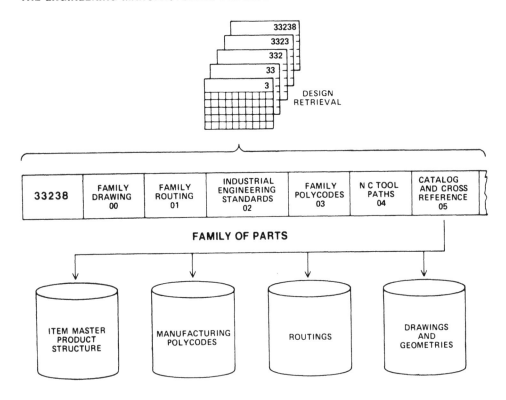

Figure 189 Shown here is a summary of the family data base reached through a classification system. Once in a family, specific parts and associated data may be obtained. The sequence of events allows an orderly storage of data in logical arrays. (Courtesy of Control Data Corporation, Minneapolis, Minn.)

weld symbols, etc., the designer can use such data to make a new drawing. It will not be necessary to correct it later. Cost penalties for changing tolerances and other standards can be defined.

An example of producibility tips is shown in Figure 190. If the designer wishes to change the tolerances of the part or increase the accuracy of the microfinish, there will be an ever-increasing cost differential. The "normal" cost of the part when the maximum tolerance is .003 in. is 100%, when .0001 in., 400%.

When realizing that data for 60,000 or more parts may be in a data base, it would seem comforting to know a retrieval system can be used to focus on just the right data needed. Assume one machine shop's operations can be defined in 14 categories:

Turning	Boring
Shaping	Threading
Drilling	Planing
Sawing	Broaching
Grinding	Hobbing

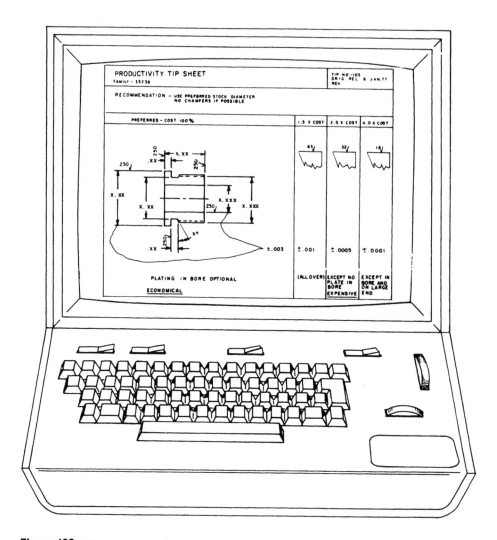

Figure 190 If we can store drawing data as a set of similar parts, we can then store design standards and drafting standards dynamically in a series of instructions to the designer called *producibility tips*. Penalties for close tolerances, finishes, etc., can be pictured for the CAD/CAM operator on the screen as shown here. (Courtesy of Control Data Corporation, Minneapolis, Minn.)

Reaming	Routing
Milling	Threading

When manufacturing has a work center data base, this data can be tied to each as a *polycode*. Thus, if all parts have individual polycodes describing which set of requirements are needed to make the part, forecasted requirements can be developed to compute machine requirements, machine load, and plant capacity. We will define a polycode:

Polycode   A set of independent sequential codes, each digit or set of digits describing individual characteristic variations. Polycodes are often described as *feature* codes.

Thus the list of features for round parts may be classified and coded, as shown in Figure 191.

These features are not necessarily a part of the design retrieval system. Since it was not necessary to use all features of all parts in the design retrieval system, the polycodes are designed to emphasize specific machining features regardless of the shape or function of the part.

As can be seen, general features, external features, internal features, and special requirements may be desired to extract and code. With this information, plans can be made to determine whether group technology is practical within a manufacturing company. Studies like the one shown in Figure 192 have proven batch sizes of parts issued to the shop are very small. The figure shows a poor example of scheduling. Imagine a company where 40% of all work orders issued to the shop are for a lot size of only one.

There have been many discussions about group technology. Relatively few companies in the United States have successfully implemented such a system. We shall explore what possibilities it has.

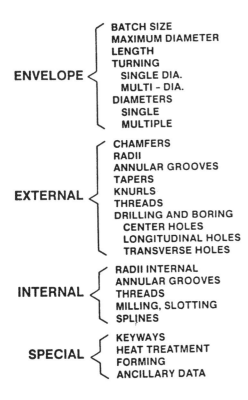

ENVELOPE
BATCH SIZE
MAXIMUM DIAMETER
LENGTH
TURNING
SINGLE DIA.
MULTI - DIA.
DIAMETERS
SINGLE
MULTIPLE

EXTERNAL
CHAMFERS
RADII
ANNULAR GROOVES
TAPERS
KNURLS
THREADS
DRILLING AND BORING
CENTER HOLES
LONGITUDINAL HOLES
TRANSVERSE HOLES

INTERNAL
RADII INTERNAL
ANNULAR GROOVES
THREADS
MILLING, SLOTTING
SPLINES

SPECIAL
KEYWAYS
HEAT TREATMENT
FORMING
ANCILLARY DATA

Figure 191   This list of typical polycodes are representative of the type of attributes which might help manufacturing isolate specific attributes of designs.

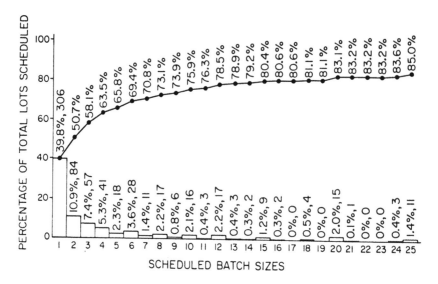

Figure 192   Most people today think of manufacturing as a large-scale mass-production system. The figures shown here disprove that idea. Scheduled batch sizes for from 1 to 17 parts in manufacturing cover 80% of all production in this example. Almost 40% of all parts are built one at a time.

Group technology is defined as follows:

Group technology   The analysis of similarly collected parts or assemblies using similar form, shape, function, or operation, such that they may be manufactured in a similar manner utilizing identical tooling or machine tools. Two distinct methods exist: Collection by type is defined as a technique utilized for similar parts using uniform data such as processes, material size, tools, and weight. Collection by group is defined as a technique where a selected group of parts have similar operations which may be combined and performed in a specialized machine shop or machine.

Figure 193 shows the types of parts which can be grouped as similar parts. Figure 194 pictures parts which can be grouped by similar operations such as boring or drilling.

One key requirement has been the need to move and group machine tools into a "cell" to optimize the system. This can be expensive. Some companies have chosen not to do this but to merely define the tools in the cell, tag them, and limit their production to group technology parts. This has proven to be quite effective and less expensive.

Without the proper disciplines of family standard processes, knowledge of part requirements, and machine tool capability, the system probably will not work very well. Similar part families grouped together may be the most promising aspect. Remember, the really small batches of parts may fall into the C class of inventory representing only 5% of the total dollars.

Remember most foremen in a machine shop attempt to eliminate excessive setups now. If group technology is to enhance manufacturing costs, it must be carefully, logically, and scientifically implemented.

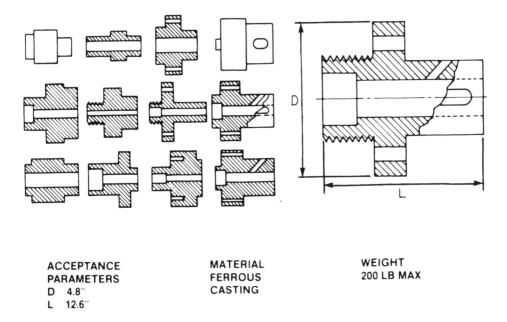

ACCEPTANCE PARAMETERS	MATERIAL	WEIGHT
D 4.8″	FERROUS	200 LB MAX
L 12.6″	CASTING	

Figure 193 Utilizing group technology, parts may be grouped together using parameters as diameter (D), length (L), material, and weight. All parts together make a composite and may be routed over the same machine tools, and together, thus increasing the number of parts made in one batch.

Figure 194 Another (and the original) concept of group technology is to bring together parts, whether similar or not, with the same operations to be performed. The parts in this picture have identical boring and drilling operations. Therefore, minimum set-up time is required.

The most promising and growing segment of manufacturing is NC tools, which will eventually encompass robotics. Manual parts programming and nonstandard post-processors have placed shackles on these tools. Now, with CAD/CAM, radical differences are appearing. By integrating the geometries of design with standard drafting practices, much of the geometric definition of the part which parts programmers require will have been done for them.

The list of applications for NC tools shown in Figure 195 has had prominently added to it the criteria of small quantities, which should have always been there.

For all the reasons listed, and more, NC tools will become more significant to productivity increases in the future. CAD/CAM systems allow a parts programmer to visually see the tool path as it is prepared. No longer are keypunching, paper-tape preparation, and two-dimensional plotting, sequential bottlenecks. Now an input trace, representing a paper tape, can be stored in the computer and changed interactively as needed. Obstructions such as part clamping can be seen and programmed over or around. The tool path can be pictured not only as a center line, but "robustly," showing its true outside dimensions. Perspective views of the part can be generated to see

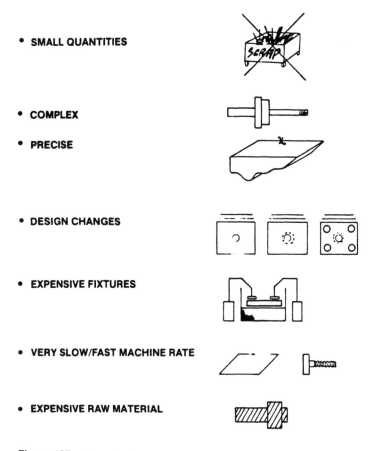

- **SMALL QUANTITIES**

- **COMPLEX**

- **PRECISE**

- **DESIGN CHANGES**

- **EXPENSIVE FIXTURES**

- **VERY SLOW/FAST MACHINE RATE**

- **EXPENSIVE RAW MATERIAL**

Figure 195   The criteria for using numerically controlled machine tools are listed here. CAD/CAM will make the first reason practical at last, manufacture of small quantities, through the "visualization" of the tool path.

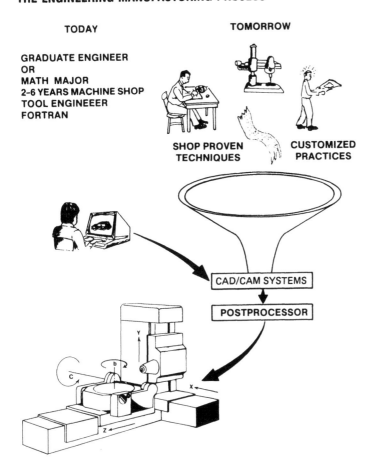

Figure 196 The highly technical personnel required today can be used to develop proven shop techniques and customized practices to reduce the technical skills necessary and thereby broaden the base of parts programmers. (Courtesy of Control Data Corporation, Minneapolis, Minn.)

if the tool shank meets interference. Pockets too small to enter with the chosen tool can be seen.

As pictured in Figure 196, these significant changes mean higher productivity with less skilled people, and many fewer test pieces made incorrectly. Job shops may become mass production centers, and will create further impetus to advance to computer numerical control, direct numerical control, or perhaps a set of automatic transfer lines.

PROCESS PLANNING

It would seem, by studying the system we are describing, a long time from release of a part to preparing a process for it. Actually, a process is prepared soon after the part release. In describing the tools available in the data base, we are now able to clearly portray how a process or routing may be prepared efficiently.

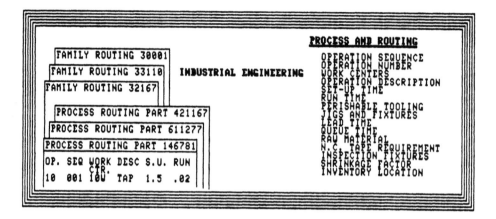

Figure 197   Preparation of routings and processes will be more rapid using the family master routing concept as the data base to alter for specific designs.

The gear box requires two new parts and one assembly. If we have family routings mechanized, the release of these new parts should automatically trigger printing of the appropriate family routings as shown in Figure 197. By scanning the drawings of the new parts, exceptions of additions or deletions to the family routings can be made, and a specific set of part routings generated immediately. What do you estimate the savings in lead times and process preparation will be?

In much the same manner, as a parts list is used to generate an engineering data base, so does process planning generate manufacturing data for shop loads, work center loads, and efficiency. All of these reports are based on data from this process.

Standard time data is often used to compute operation and set-up times as opposed to a stopwatch. General empirical rules are used to develop queue time between operations. These are important, for they become part of the lead time for material requirements planning as well as for production scheduling. The proper tools and fixtures are designated here, and new tooling may be required. Raw material for fabricated parts, including shearing and/or burning sequences are also determined. It should not be forgotten that, prior to any of this work, a make or buy decision was also investigated.

The industrial engineer is an important part of a manufacturing system. Unfortunately, many companies have none, or too few. Some of the tools mentioned in this booklet may serve to alleviate the shortage.

PURCHASING

If the data base is properly oriented, purchase orders, or, at least, requisitions should be printed automatically. Buyer and vendor codes may be used for automatic distribution and load leveling on the purchasing agent, dividing the parts into similar groups so they can specialize by types of parts and/or by vendors. Other reports assist purchasing also. Some of them are shown in Figure 198.

Too often the best MRP system is spoiled because of late vendor deliveries. Often the vendor is not aware of this shortcoming. By recording and tabulating promised versus actual receiving dates, and notifying vendors of delinquency, safety stocks can be minimized.

Many times companies have more than one manufacturing plant with their own buyers. By combining purchases of all plants into a coordinated report it is possible to see whether the same part is being purchased for more than one price. Standard versus actual costs, cost trends, and other allied data will assist in managing purchasing costs which may well represent 50% of the total product cost. Too little emphasis has been placed in the past on the role a mechanized purchasing system can perform in the overall manufacturing efficiency.

A mechanized purchasing system is most important to an on-line receiving system. Immediately a purchase order is written it is then possible to receive the material. Thus, timely inventory status becomes the responsibility of purchasing.

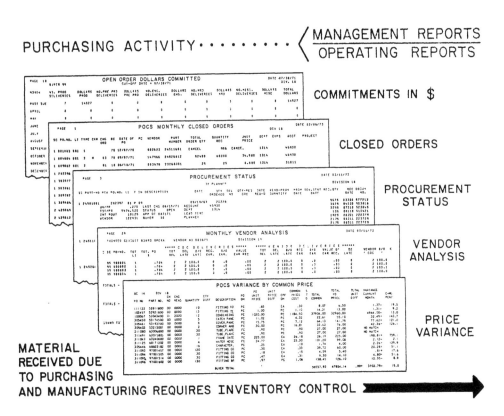

Figure 198 The purchasing department gains by mechanizing their procedures. Some of the data available as a result are indicated with these reports.

TOOLING

CAD/CAM may become the most important system to increase the productivity in designing and building tools. It seems that many difficulties must be met when designing a piece of equipment round a hole, yet the tool designer often does. The drawings of "simple" tools for the machine shop are shown in Figure 199. If we can picture a screen showing the actual part to scale, and then superimposing the tool around it, and erasing a part whenever necessary, tool design should be greatly simplified. In addition forging dies, and casting molds require the same attention as a hole. It would seem that a CAD/CAM system, having an ability to make an inverse part, would assist in designing the die and the mold. We must begin to determine the economics of replacing wood patterns with aluminum machined directly from the inverse design of the part on CAD/CAM. In the same way, combining the mathematics of progressive forging dies with the inverse mathematics of a part should create the forging die. Thus, the tooling which cannot be replaced using a numerically controlled machine tool should be prepared using one.

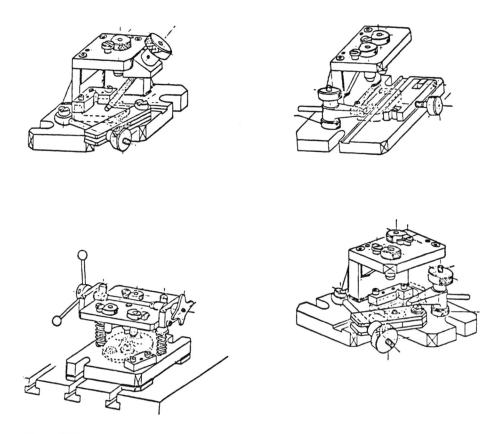

Figure 199 Similar, even identical, tooling can be found if tied to a classification system, thus eliminating months of lead time.

PRODUCTION SCHEDULE

The production schedule is an amalgamation of many systems. The forecast and bill of material have generated material requirements. The process has developed the potential machine tool requirements and queue times, and inventory control has determined the next requirements. The production schedule represents the best knowledge of physical requirements of parts for each time period. Usually the production schedule is the result of an "infinite" load. No recognition that some particular work center is overloaded was taken into account during its generation. If a finite load is computed, *string loading,* and priorities must be developed. Under these rules, movement of a part requirement backward in time is used to reduce the load on a center. However, if one part of an assembly is moved, the others are not needed as soon, and should also be moved. Any reloading as described is complex, since a reschedule can alleviate one work center and load another. Several different reports to assist production control are shown in Figure 200.

Many algorithms have been tried, including, linear programming techniques, but have been generally ineffective. The combinations and permutations are, for now, too complex. It is also difficult to determine how to set priorities. Should efficiency of the operator, or throughput of parts be the criteria?

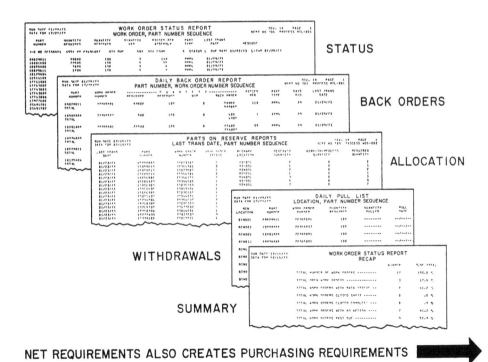

Figure 200 Production control reports indicated here are the result of a sound mechanized systems whose roots stem from accurate engineering documentation.

In the final analysis, a considerable responsibility for on-time production falls into the hands of foremen, supervisors, and expeditors, those air-traffic controllers of the factory aisles. Thus, *hot lists* replace orderly production schedules in many plants. Priorities may change hourly.

Even under these conditions, more and more production scheduling has become an effective force in the factory. As large-scale computers and their data bases are used to apply more scientific methods, and on-line systems become a part of the foremen's tools, the production scheduling techniques will become more meaningful, accurate, and useable.

SHOP FLOOR CONTROL

Two shop packets are shown in Figure 201, one for a piece part, and one for an assembly. Each packet is prepared with a drawing and a process. The process usually covers two requirements since it may also be printed as a work order showing quantities required. Either punched cards, or perhaps an on-line data collection system is used to report labor during the operations being run. Additional punched cards, or some other alternate is used to requisition material from the storage yard.

For an assembly, an additional piece of paper known as a *picking list* is required.

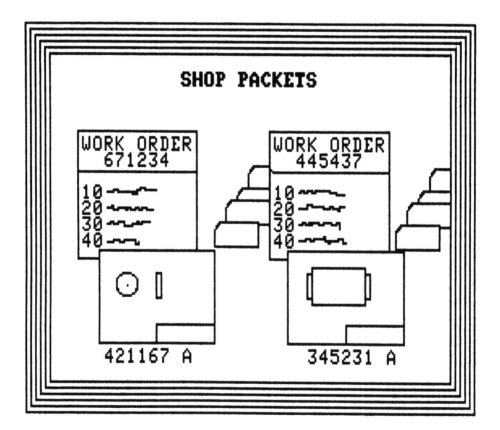

Figure 201   Shop packets are used to notify the shop floor of work required. The routing, drawings, and, often, time cards are in the packet.

The picking list is first sent to inventory control at a predetermined time prior to assembly to select the parts needed for the assembly. Usually the picking list is printed in bin location sequence, which saves up to 50% of the time locating and staging the parts. As can be seen, the proper structure of the parts to make an assembly is extremely important otherwise excessive parts will be picked, or inadequate parts will be issued.

The picking list represents the present configuration of the product. Engineering may have changed the design several times and be up to revision level 10 while manufacturing is still assembling to revision 6. The need for configuration control is graphically illustrated should manufacturing have no confidence in this picking list. Lost time calling engineering for confirmation is expensive.

Several lead times are involved in the shop packet. Preparation time, picking time, staging time all may be required to issue the shop packet on a timely basis. Too often shop packets are issued prematurely, thus loading the shop unnecessarily and causing "cavitation" in the pipe line of production.

INSPECTION

Inspection has often erroneously been thought of as a necessary evil. Today foreign competition has proven the error of this philosophy. The picture of a face plate and height gauge in Figure 202 is typical in today's manufacturing. Often one part of a batch is extracted from production and inspected. First-piece inspection before a long manufacturing run is also usual. Face plates and gauges will eventually be replaced by electronic inspection fixtures. It is important that we pursue them to speed inspection and increase its accuracy. In World War II, quality control became the most efficient means of controlling quality. "X" bar, "R" charts were to be found at each machine tool and samples of each part's dimensions were recorded and statistically tested. Today tolerances are becoming much tighter, and human inspection, even with the best of tools, is much more difficult and time consuming.

We must also recognize a serious potential problem. Numerically controlled (NC) machine tools can now hold tolerances in the millionths of inches. We must not allow designers to use these tolerances just because they are possible. It is important to begin simulating assembly tolerances necessary to judge them before they are applied to parts. Today simulation techniques on a computer can be used to perform this function.

GO, NO–GO gauges are not efficient to obtain quality parts, and optical comparison tools, while accurate, are not fast enough. The entire field of inspection needs rejuvenation and revision, and computer-aided design/computer-aided manufacturing (CAD/CAM) techniques seem to be the most promising for the future.

MACHINE SHOP

Briefly we must look at the actual production. The machine shop or the fabrication shop is a compartmentalized, segmented, but powerful machine. However, new studies have not been made to determine optimum location of machine tool, traffic patterns, optimum storage, tool-rack locations, perishable tool standardization, tool location systems, and a host of others. It seems as though we consider modernization of each facet of manufacturing one step at a time.

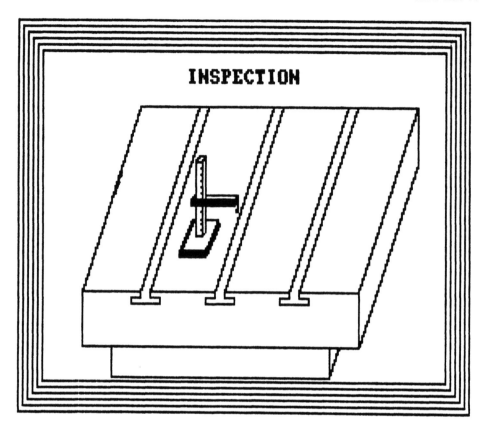

Figure 202 Inspection is slowly converting from this type of equipment to electronic measuring devices which will increase the speed and accuracy.

Integration of numerically controlled machine tools as pictured in Figure 203, and their parts programmers should be studied. Too often the parts programmers are partitioned in a separate booth apart from the action. Greasy files of old prints, and parts which were issued and never used litter the floors, aisles, and every extra corner. Hundreds of parts sit gathering dust or rust. Trees grow through old castings. Grease covers the aisles.

The companies who do not have these conditions are numerous. For those who do have some of the problems mentioned here, simulations of new traffic patterns, new machine locations, better storage bin allocations, surplus inventory evaluation, standard tools, all will be helpful to tighten up the machine shop and make it more efficient. If productivity is to be increased, solving these problems is as important as any new system.

ASSEMBLY

An assembly floor is composed of a variety of work methods. For mass production, mechanized assembly-line loading programs, loading people per station, and optimizing production quantity are used. Linear programming techniques optimize product mix. For large equipment, the largest part of the equipment is set in place and the rest of the parts are brought to it. The moving conveyer is used for assemblying toasters, vacuum cleaners, and tractors. In a job shop, the final test of a product is whether it will go together.

Most of the engineering changes occur at assembly. Thus, CAD/CAM will become a significant tool in simulating assemblies long before they reach that stage. Physical prototypes have been made in the past to iron out the "bugs" of assembly, but this is usually a luxury only those who mass produce can afford. The job shops rely on adequate design for each order. Thus, we will see a great emphasis placed on modelling of finished products by such companies when they see the benefits to be derived from CAD/CAM. Too often accounting requirements slow down efforts to improve the efficiency of manufacturing. Insistence on keeping track of parts and assemblies by customer order numbers eliminates the possibility for grouping like parts and assemblies in manufacturing. Too often those who insist on such segregation do not recognize the similarities between jobs that could generate the efficiencies of mass production. By utilizing all the tools mentioned in this booklet, we will see a significant shift from accounting-oriented rules to more practical and efficient methods.

Robotics will begin to make inroads on the assembly process only when the product is redesigned to use them. Too many complex assemblies exist. Not only will robots require improved techniques, but designs will have to be studied with robot abilities constantly examined.

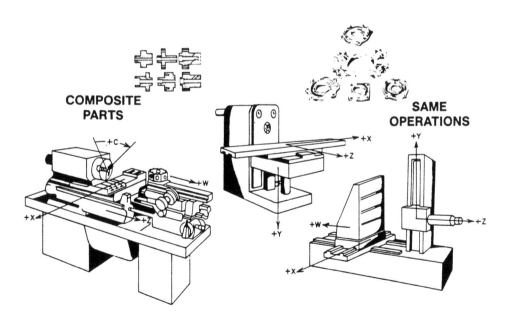

Figure 203 Slowly but surely old-fashioned machine tools are being replaced with numerically controlled machine tools and robots.

CONCLUSION

If we can summarize by "putting it all together," Figure 204 expresses what we have tried to convey in this chapter.

Utilizing an integrated system will pay large dividends. By taking a forecast and using it to explode product bills of material, we can obtain total part and assembly

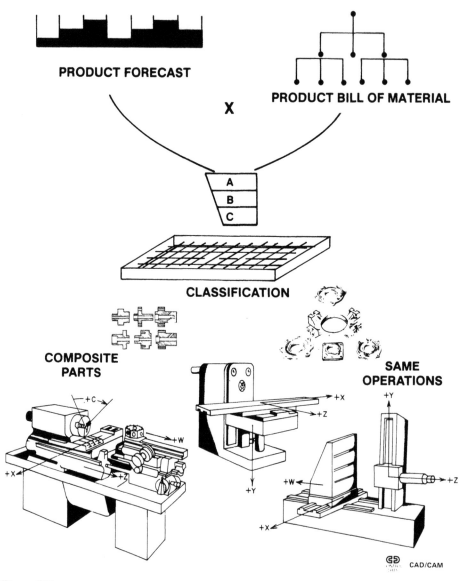

Figure 204 By integrating business and CAD/CAM systems, the slash can be removed from CAD/CAM. Combining the product forecast with the product bills of material generates requirements. Sorting the requirements by usage and by family types develops production requirements for group technology cells or mass production. A machine capacity forecast is another result. (Courtesy of Control Data Corporation, Minneapolis, Minn.)

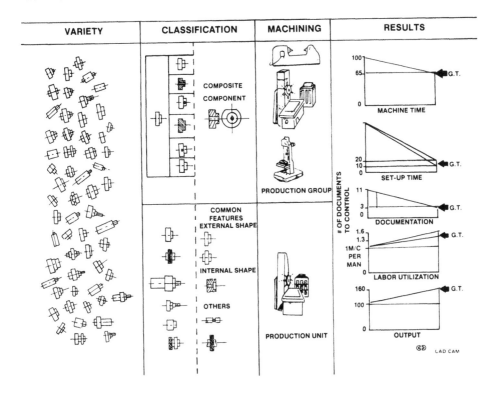

Figure 205 Manufacturing systems integration has been satisfactorily achieved, even though manually. One company classified a variety of parts to develop composite components and those with similar shape or features. Production groups or units of machine tools were prepared. The result was a significant decrease in machining time, set-up time, documentation paper work requirements, with a corresponding increase in labor utilization and product output. (Courtesy of Control Data Corporation, Minneapolis, Minn.)

requirements for the future. By sorting this data through an inventory analysis, we can then convert our results, through a classification system, into determining the type of machine tools required, the cells of machine tools or production cells needed for group technology, and finally, predict the deficiencies, the surpluses, and the obsolete, all in a timely, logical, and economical way.

One company who tried these methods reported the following results, as shown in Figure 205:

1. Machine time decreased 65%
2. Set-up time decreased 80%
3. Documents required 3 of 11
4. Man–machine ratio 3 : 6
5. Output increased by 160%

THE FUTURE

Figure 206 is an encapsulation of what CAD/CAM really means. Without question CAD/CAM has broadened the horizons of engineering and manufacturing. It will create a friendly revolution in the art and science in the design and manufacture of products and goods.

Utilizing the advanced mathematics we know today, and enlisting all the power of a large-scale computer and its associated data base, such esoteric disciplines as geometric modelling, design analysis, kinematics, and automated drafting will be used to form the most powerful set of data ever prepared.

From the data base thus generated, further action data will be added to determine the most effective means for producing parts and assemblies of the product utilizing numerically controlled machine tools, robotics, process planning of part families, and direct numerical control, thereby increasing the ability and efficiency of design and factory management.

The Automated Factory Is Here

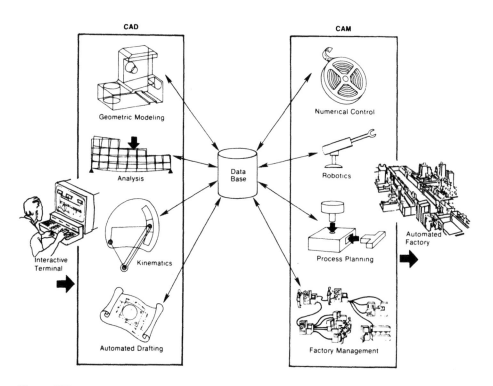

Figure 206   By utilizing a data base concept, CAD/CAM, through interactive terminals, may be used to enter geometric modeling data, analysis, kenematics, and factory management data. The next step will be an automated factory. (Courtesy of *Machine Design Magazine,* Cleveland, Ohio.)

A FINAL LOOK AT MECHANIZED SYSTEMS

12

Years of data processing have not always been profitable to many businesses. Engineering and manufacturing systems have created almost as many problems as they have attempted to solve. Millions of dollars have been spent in an attempt to make manufacturing more economical and streamlined. So what may have gone wrong?

For many companies the history of data processing can be summarized in seven brief statements. Implementation of a mechanized system has followed the path outlined by them:

Unwarranted Enthusiasm
Uncritical Acceptance
Growing Concern
Unmitigated Disaster
Search For The Guilty
Punish The Innocent
Promote The Uninvolved

During this period many managers of data processing became the "wandering gypsies" of the western world. Ordered to perform rapidly (or volunteering), data processing managers implemented systems before users were able to comprehend and prepare for the radical changes and increased disciplines necessary.

When visiting almost any manufacturing company the "system" employed is usually a series of individual programs or activities which make use of the computer. Many companies have such systems as order processing, production-line monitering, materials flow control, shipping and receiving, and others. They simply do not talk to each other. Many systems have been designed so the user works for the computer rather than the opposite.

What has gone wrong? Some possibilities are

The system may have been designed for the computer and not the user. It may have required a computer size not available.

The total system requirements may have been ignored for more rapid implementation.

Often poor initiation, maintenance, and education have been at fault.

A crisp specification has often been missing leading to "glittering generalities."

With no imagination, it is quite possible to make a computer system work more inefficiently than the original manual system.

Perhaps the most significant problem has been the lack of management participation. In many cases they abdicated their responsibilities by turning over the job to young, ambitious, but green systems analysts. Yes, they approved spending money for new systems, but they have had little understanding, and taken little personal action to see them operate effectively.

It is so easy to overdesign a system. If one problem is solved, it is easy to expand to another and another, until a relatively simple system becomes too complex to handle. Many such are still with us!

There are a few companies who have managed to implement systems with relatively good to excellent results. These companies may take pride. It is their success which has continued to hold out hope for others. We should be advised that placing the system requirements ahead of all other considerations made them successful.

We should not forget that it is people that make a system operate. We must recognize their shortcomings to help them achieve success. Some of these characteristics are listed here:

EAGER, but unaware of other people's systems

ORIENTED to parochial systems

UNAWARE of advanced methods

UNABLE to analyze their problems objectively

LIMITED in education

VALUABLE only in their present job

FRUSTRATED in their desire to learn more

Manufacturing companies need a new approach, a new formula, a new recipe to achieve the success which has eluded them.

More training and education is a significant, but not total corrective measure. Management interest, participation, and active responsibility has become mandatory. They can be assisted by a task force, utilizing the best persons the company has to offer. Recognition that it takes three to five years to implement a sound and economical integrated system will help also. Perhaps this book has helped demonstrate that the study and successful design of engineering documentation is a significant part of this activity.

INDEX

9 780824 770891